THE BODY AND EVERYDAY LIFE

In recent years, there has been an explosion of interest in the contemporary social study of the body which has raised important theoretical and methodological questions regarding traditional social and cultural analysis. It has also generated corporeal theories that highlight the fluid, shifting, yet situated character of the body in society. In turn, these corporeal theories have implications for social relations in an era of new technologies and global market economies.

The Body and Everyday Life offers a lively and comprehensive introduction to the study of the body. It uses case studies in performance practices to examine the key concepts, methods and critical insights gained from this area. It includes sections on:

- ethnographies of the body
- bodies of performance
- performing gender
- the ageing performing body.

This book clearly illustrates the complex relationships that exist between the body, society and everyday life, and considers the negative and positive implications for the development of future socio-cultural analysis in the field. It will be an invaluable introduction for students of sociology, body studies, gender studies, dance and performance, and cultural studies.

Helen Thomas is Director of Doctoral Programmes at the University of the Arts London. Trained in both dance and sociology, her research interests centre on the sociology of dance and the body, cultural theory, and qualitative research methods applied to cultural practices. She has published numerous articles and book chapters and the following books: *The Body, Dance and Cultural Theory* (2003); *Cultural Bodies* (eds. with J. Ahmed, 2004); *Dance in the City* (ed. 1997), *Dance, Modernity and Culture* (1995) and *Dance, Gender and Culture* (ed. 1993).

THE NEW SOCIOLOGY

Series Editor: ANTHONY ELLIOTT, Director of the Hawke Research Institute, University of South Australia

The New Sociology is a book series designed to introduce students to new issues and themes in social sciences today. What makes the series distinctive, as compared with other competing introductory textbooks, is a strong emphasis not just on key concepts and ideas but on how these play out in everyday life – on how theories and concepts are lived at the level of selfhood and cultural identities, how they are embedded in interpersonal relationships, and how they are shaped by, and shape, broader social processes.

Titles in the series:

Religion and Everyday Life
STEPHEN HUNT

Culture and Everyday Life
DAVID INGLIS

Consumption and Everyday Life
MARK PATERSON

Community and Everyday Life
GRAHAM DAY

Ethnicity and Everyday Life
CHRISTIAN KARNER

Globalization and Everyday Life
LARRY RAY

Gender and Everyday Life
MARY HOLMES

Self-Identity and Everyday Life
HARVIE FERGUSON

Risk, Vulnerability and Everyday Life
IAIN WILKINSON

The Body and Everyday Life
HELEN THOMAS

Forthcoming titles in the series:

Cities and Everyday Life
DAVID PARKER

Nationalism and Everyday Life
JANE HINDLEY

Media and Everyday Life
ELLIS CASHMORE

THE BODY AND EVERYDAY LIFE

HELEN THOMAS

LONDON AND NEW YORK

First published 2013
by Routledge
2 Park Square, Milton Park, Abingdon, Oxon OX14 4RN

Simultaneously published in the USA and Canada
by Routledge
711 Third Avenue, New York, NY 10017

Routledge is an imprint of the Taylor & Francis Group, an informa business

British Library Cataloguing in Publication Data
A catalogue record for this book is available from the British Library

Library of Congress Cataloging in Publication Data
Thomas, Helen, 1947–
The body and everyday life / Helen Thomas.
p. cm. — (The new sociology)
Includes bibliographical references and index.
1. Human body—Social aspects. 2. Human body in popular culture. 3. Human figure in art. 4. Body image. I. Title.
HM636.T475 2013
306.4′613—dc23

2012036249

ISBN: 978-0-415-33111-1 (hbk)
ISBN: 978-0-415-33112-8 (pbk)
ISBN: 978-0-203-39234-8 (ebk)

Typeset in Garamond MT
by RefineCatch Limited, Bungay, Suffolk

Printed and bound in Great Britain by MPG Printgroup

For David and Paul

CONTENTS

ACKNOWLEDGEMENTS

This book has been in the process of 'becoming' for a long time but for one reason or another I was unable to complete it until now, although three of the five chapters had been in full draft for a number of years before this. I would like to thank Routledge for not giving up on it or me and especially Emily Briggs, the current Editorial Assistant, for pushing me to completion in the last year with regular email reminders for updates on final submission. Fellow academics and dear friends, Gay Morris, Stacey Prickett and Jen Tarr took time out of their busy work schedules to read the manuscript and with their usual critical attention to detail and sense making, provided me with comments, questions for clarification, expansion etc. for which I am extremely grateful. Jen in particular went through the document with a fine tooth comb and offered a number of suggestions for improvement, as well as correcting some glaring typos. Thanks, too, to Sarah Carter who proof-read the book in her own time. In the final chapter I draw directly from a report, *Dancing into the Third Age: Social Dance as Cultural Text* (2003), authored by Lesley Cooper and me and I gratefully acknowledge the contribution Lesley made to this. Despite all this help and support, the final responsibility for the contents lies entirely with me.

1

INTRODUCTION: ENTER 'THE BODY'

Most people tend to take their body for granted as a 'natural' fact of existence without thinking about it too much on a moment-to-moment basis. In western culture, the body is construed to be a fairly reliable instrument through which people express and represent their conscious self, their individuality, in their everyday life. The tendency is not to question or think about bodily actions in any detail as individuals go about their daily business; the body and the self, for the most part, seem to operate in a relatively stable relationship to each other, a kind of unspoken mind/body integration, in which the conscious, willed self is considered to have the upper-hand.

Needless to say, there are also numerous occasions when this taken for granted integration is momentarily disrupted and the body, yours and others, is called into question. This is increasingly so with the current preoccupation with and on the body in contemporary consumer culture. On a fairly mundane level, think of the shock of seeing yourself in a mirror when you are not expecting it. The passing vague awareness that you are observing another person's image turns into instant horror when you realize that it is you in that mirror and you never thought you looked quite like that. At which point you attempt to adjust your 'look' to make it more in keeping with the bodily image you have of/for yourself in your head, or the one you put on when you 'intentionally' look in the mirror, or that which you wish to display to/for others. Similarly, images of other people's bodies (usually of a fairly uniform kind), in various dressed, semi-dressed, undressed states, which are constantly on display on billboards, in the tabloids, magazines, television and film, act as a reminder to us of the observed differences and similarities between the 'touched up', often idealized, bodies on display before our eyes and our individual 'ordinary', perceptual, 'real', usually more fleshy, flawed body. Whilst women have traditionally been the subject and object of such representations, and there is evidence to show that young women today feel more free to play with these (Crossley 2006), men have also been recently brought into the ubiquitous circulatory economy of bodily images (Featherstone 1991; Goldstein 1994; Bordo 1994; Davis 2003).

On a more serious and perhaps more sustained note, the customary out-of-awareness body is also disrupted when, through illness, ageing, injury, trauma or disability, the body does not perform as it is expected to, or as it has habitually done in the past. For example, sometime before his death, the septuagenarian author John Mortimer (2001) wrote movingly of the day he realized that he could no longer bend down and put on his socks as a moment of stark recognition of his impending old age, bodily decay and mortality. In *The Scar of Visibility*, Petra Kuppers, a cultural disability activist, feminist and scholar of performance studies, notes that her methodological approach is grounded on 'embodied acts of perception' which entail 'attention of corporal acts of meaning making, coding and decoding' (2007: 3). Drawing on the work of cultural theorist Walter Benjamin, Kuppers approaches her study of medical performances and contemporary art as a 'flâneur of bodily creative practices' (ibid.). Benjamin's notion of the flâneur, as Kuppers notes, 'is always male and moves with disinterest and non-attention to movement around the spaces of modernity' (ibid.). The flâneur's practice of bracketing out attention and interest in movement, however, has never been available to her because, as a disabled woman, attention and interest in movement are a necessary and consistent part of her everyday life. Hence, Kuppers' 'productive intervention' into the domain of the flâneur challenges the idea that 'the body' usually remains out of awareness in everyday life, rather, it is perhaps not the case for certain social groupings and 'different bodies'.

Wendy Seymour's (1998: xiii) qualitative analysis of people who have 'experienced profound permanent bodily paralysis' also shows how severe physical impairment disrupts the sense of self. In such instances, the routine stable relationship of the self and the body is thrown off balance and the body seems to take on the character of an external or foreign object, which demands attention. It is no longer perceived as an instrument of the mind or human agency. A crisis arises between the body and the mind, in which the objectified body seems to have a mind of its own, which demands attention; it takes on a 'thing-like' status, which the individual concerned becomes acutely aware of and attentive to, the subtlest, smallest changes in its condition. Seymour does not restrict her analysis to the negative ramifications of the assault on the taken for granted self-body integration which occurs as a consequence of sudden, major physical impairment. Rather, her study explores the processes involved in 'remaking the body' because, as she argues, it is impossible to have a self without a body. At the same time, the study proposes, as others have also argued (see Evans and Lee 2002) that 'real' bodies/selves are not fixed or immutable.

There are areas of work or activity in everyday life, of course, which require significant attentiveness to the body and which, over time, seem to become 'naturalized' in the habitual everyday behaviour of the individuals involved. Dancers, for example, generally have what appears to be a 'natural' attentiveness to the body (see also, Loïc Waquant's [1995, 1998] ethnographic study of professional boxers). They seem to move in more precise and considered ways than the bulk of the population, even as they walk down the street, perhaps as a consequence of their rigorous 'bodily' training regimes which are designed to hone and transform the body. In her study of *sinulog* dance forms in the Philippines, Sally Ann Ness (1992) notes that her training as a dancer/choreographer often impacts on her everyday life in a way where mundane tasks, which she terms 'taskless tasks', like folding a towel for example, take on the character of a dance. In such instances, Ness senses that she performs the task so much 'better' than is necessary or usual:

> The rhythm of the arm movements is more articulately phrased than it needs to be, the sections of the towel come out to be more equally divided than they really need to be, the contact of my fingers with the fabric of the towels is more delicate than it needs to be, and so forth. These characteristics begin to turn the ordinary action into choreography . . .
>
> (Ness 1992: 7)

The overwhelming majority of professional performers in western theatrical dance are young, predominately female and thin, with bodies that have to be able to master the ever-increasing technical challenges that specific dance training regimes, choreographers and performance practices require of them and which are predicated on the idea of an abstract body without 'impairment' and/or 'disability'. For a discussion of the debates around the use of terms such as 'impairment' and 'disability' in disability studies see Tom Shakespeare and Nicholas Watson's article on 'The Social Model of Disability' (2001). However, although in the minority, it is also the case that disabled performers, as with some of the dancers in the modern dance company, CandoCo, which is a company of disabled and non-disabled dancers, can and do demonstrate the remarkable possibilities of moving bodies both on and off the stage, regardless of and to an extent because of, what are generally considered as specific (physical) 'impairments'. Indeed, it can be argued that wheelchairs and partners, for example, extend the individual body in new relational ways, opening up new possibilities for performance from both the performer's and the audience's point of view.

The restaurant trade provides a heightened, although often unnoticed, concern with bodily matters, in addition to the more obvious concerns with hygiene. Waiters and kitchen staff, often carrying plates, implements or hot pans, which function as bodily extensions, have to learn to move around each other in confined spaces with great economy, spatial precision and timing, if they are to prevent bodily clashes, accidents and food disasters, or the wrath of the chef. Elspeth Probyn (2004), for instance, describes to great effect the choreographic-like practices of the front of house and kitchen staff in a restaurant in Québec where she worked while she was a student. These involved the staff, including herself, dancing around each other, through kitchen doors, between tables and customers, with consummate ease, economy of movement and at breakneck speed, to someone else's timing, usually the chef's. That is, when it all came together and 'you were just on the verge':

> That's when everything flowed: your drink orders were ready, then the appetizers were quickly eaten, the plates cleared just as some inner clock told you the mains were up. On huge bus-trays you could easily stack a couple of tables of grub. Then with a heft to your shoulder, a kick to the kitchen door you'd be out, meals delivered and back again for the next. Along the way making eye contact and a promise to be back to the new tables, you'd direct the busboy to wipe down an empty table, while you deposited bread on the table.
>
> (Probyn 2004: 222–3)

This attentiveness to the body, encapsulated in these few examples, the recognition that we are embodied beings, that we not only 'have' bodies which are enabling and delimiting, but that to a certain extent we 'are' bodies connected to other bodies, as Bryan Turner (1984) has noted, and that some are more privileged than others, is precisely what inspires this text on the sociology of the body. In this book, the 'not everyday' is used to shed light on everyday bodies. By drawing on a range of performative practices which include dancing, live art, documentary photography and boxing, I hope to cast light on the complex relations of the body in the social world and *vice versa*, raising questions along the way as to 'what is a body?' (Fraser and Greco 2005). The answers, as will become evident in the course of the book, depend on the approach that is taken by the social or cultural analyst.

BODIES – THEN AND NOW

If you are studying sociology or a related discipline at undergraduate or postgraduate level, you will be aware that a great deal of work has been published which has the word 'body' in the title or sub-title over the past 20 or so years. This fascination with the body in sociology, at least in a systematic way, however, is relatively new in terms of the development of the sociological tradition.

Whilst working on my PhD on the sociology of dance in the late 1970s and early 1980s, for example, I sought out sociological and anthropological sources that included an interest in 'the body' in order to critically review the literature. The body is heavily implicated in dance, particularly in the west, where it largely constitutes the means of expression and mode of representation. As there was a paucity of material on dance or dancing from a sociological perspective at that time, I imagined that I might gain some sociological insight by seeking out studies on the body. It soon became apparent, however, that the body was not exactly a 'hot topic' at that moment in sociology, any more than was dance. Instead, my attention was directed towards certain interdisciplinary 'behaviourist'-based studies which sought to generate a serious 'scientific' approach to studying body language in the US (Birdwhistell 1973; Hall 1969; Scheflen 1964); social constructionist approaches to the body as a medium of expression in the UK (Benthall and Polhemus 1975; Polhemus 1978); Durkheimian-inspired anthropological work on body symbolism (Mauss 1973, Douglas 1970, 1973; Needham 1973), and the occasional one or two collections which explored the issues confronting the anthropology of the body from the vantage point of psychology, anthropology and ethology (Blacking 1977; Polhemus 1978). The burgeoning second-wave feminist movement in the 1970s, which directed attention towards women's bodies as a site of political contestation, identification and representation (see Bordo 1993), also warranted my close attention.

Much of this work, apart from the body symbolism approach, as Ted Polhemus (1978) noted at the time, along with feminist scholarship, remained marginal to sociology. What was important sociologically about the Durkheimian-inspired work is that it pointed to 'the body' as a legitimate topic of sociological investigation and sought to provide a method for analysing the relations between the body and society. I will return to the impact of feminist scholarship later in the chapter as it could be argued that feminists blazed the trail that led to sociology treating the body as a serious topic of inquiry (see Frank 1991; Davis 1997).

A few sociologists, such as Harold Garfinkel (1984 [1967]) and David Sudnow (1993 [1978]), did raise questions concerning the body in their research, drawing on the influence of social phenomenology. Erving Goffman, the maverick 'interactionist' sociologist of the 1960s and 1970s, produced remarkable insights into the ways we present ourselves to others in everyday life (1971 [1959]). He also explored our routine negotiation of the unspoken and for the most part, out-of-awareness, rules of appropriate bodily behaviour in public places (1963, 1972). In this work, he combined the insights of Ray Birdwhistell's (1973) study of kinesics (everyday body movement) and Edward Hall's (1969) study of proxemics (the comparative study of interpersonal spatial relations), with a strong Durkheimian concern with societal rules and norms. Although these studies were sporadic in that period, in retrospect they perhaps hinted at something that was about to appear on the academic horizon.

Fast-forward twenty years and the socioscape with regard to the study of the body looked remarkably different. As the 1980s progressed, evidence of a more sustained interest in the body began to emerge, through key 'body books' such as Bryan Turner's *Body and Society* (1984) and John O'Neill's *Five Bodies* (1985).

Since the late-1980s, hard on the heels of what has come to be known as the 'cultural turn' in sociology (Chaney 1994), sociologists and cultural analysts have become increasingly interested in exploring and reviewing the complex relations of 'the body' in society and culture from a variety of perspectives and thematic concerns (see for example, Featherstone *et al.* 1991; Synnott 1993; Shilling 1993; Morgan and Scott 1993; Crossley 1994, 2001a; Davis 1997, 2003; Williams and Bendelow 1998; Nettleton and Watson 1998; Burkitt 1999; Evans and Lee 2002; Coupland and Gwyn 2003).

This interest has been evidenced in a range of sub-disciplines of sociology, such as the sociology of health and illness (Nettleton and Watson 1998; Watson and Cunningham-Burley 2001) and the increasingly interdisciplinary subject field of 'body studies', perhaps best exemplified by the journal *Body and Society*, which was instituted in 1995 by founding editors Mike Featherstone and Bryan Turner. But interest has also abounded in other areas such as feminist scholarship, which brought to bear new ways of thinking about embodied subjectivities, drawing on a range of theoretical frames from poststructuralism, phenomenology and psychoanalysis (Martin 1987; Butler 1990, 1993; Grosz 1993, 1994; Grosz and Probyn 1995; Davis 1995, 1997, 2003; Gatens 1996; Shildrick 1997; Price and Shildrick 1999; Frost 2001; McKie and Backett-Milburn 2001; Aaron 2001; Fraser 2003).

The burgeoning field of performance/dance studies provided fruitful ground for exploring the body in performance, particularly around the area of gender and sexuality (Phelan 1993; Goellner and Murphy 1995; Koritz 1995; Burt 1995, 1998; Foster 1996; Schneider 1997; Desmond 1997). Anthropology, which has demonstrated an interest historically in 'other' bodies in particular, also began to apply different theoretical tools to study the body (Csordas 1994, 2002; Martin 1994; Hastrup 1995). Although philosophy, like anthropology, had not ignored the body, it tended to see the body in negative terms. The work of philosophers such as Lingis (1994); Welton (1998, 1999); Weiss and Haber (1999); Weiss (1999); Punday (2003) and Shusterman (2008) sought to get away from the negative connotations and reinstate the somatic into their work. Cultural studies also took to exploring the body in the context of the media, culture and society (Gaines and Herzog 1990; Goldstein 1991, 1994; Gamman and Makinen 1994; Falk 1994; Desmond 1997; Entwistle 2000; Entwistle and Wilson 2001: Cook *et al.* 2003; Pitts 2003; Thomas and Ahmed 2004).

While artists have traditionally paid a great deal of attention to the body and its expression, recent visual culture analysis has challenged the all too easy celebration of the 'beautiful' body in western art which is revealed to be underscored by objectification in terms of gender and race in particular (Pollock 1988; Nead 1992; Betterton 1987, 1996; Curti 1998; Gilman 1995, 1999). The topic of the body has also strayed into the domain of social geography (Pile 1996; Duncan 1996; Nast and Pile 1998).

The aforementioned references represent a small section of the many monographs, edited collections and readers which came to have the body as their focus. Indeed, it is fair to say that the study of the body turned into a significant academic cultural industry, if the number of conferences, articles, collections and books which have focused on the body in social and cultural studies in the past 20 years or so are anything to go by. Moreover, in 2003, as if to give the 'somatic turn' in sociology the legitimate seal of approval, Routledge published a four-volume library edition entitled *The Body* in its Critical Concepts of Sociology series, with 2,888 pages, which would have been unthinkable 13 years before.

The first thing we might want to ask is, why did sociology show little interest in studying the body until relatively recently? I do not intend to offer a detailed discussion of this for fear of not getting off the starting blocks and because it is something which is not quite as simple as first appears. There are a number of key texts that address this in a more sustained and systematic fashion (B. Turner 1991; Frank 1990, 1991; Shilling 1993; Williams and Bendelow 1998; Burkitt 1999) than is possible here. Rather, in the following section, I will offer a gloss of the reasoning behind this seeming lack of attention in the sociological tradition. As I hope to show in the course of this book, this apparent absence has impacted on new developments in rather surprising ways. After this brief 'historical' discussion, this introductory chapter will address the explanations that have been offered for the recent surge of interest in 'the body', which was thrust on to the intellectual map in the late twentieth century. I will also explore key issues that emerge from various attempts to bring the body back into the light. Again, as will be shown in the course of the book, these often reflect back on the very concerns that writers sought to overcome in the first instance.

THE NEGLECTED BODY PROBLEM IN THE SOCIOLOGICAL TRADITION

Sociology, it has been argued, did not generate a thoroughgoing interest in the body until relatively recently because the discipline had largely been premised on the Cartesian heritage, which advocated a split between the mind and the body (B. Turner 1984; Howson and Inglis 2001). The mind, in Cartesian thought, was not only perceived to be distinct from the body, it was also superior to it. The former was deemed to be the world of culture, rationality and action, and the latter of nature or inert matter; 'flesh' or 'meat', as Drew Leder (1990) has put it. In order to establish itself as an independent 'scientific' discipline in the late nineteenth century, sociology needed to have its own object of study and rules of investigation, which were different from and not reducible to those of psychology, anthropology or the natural sciences. The clearest statement of this can be found in Emile Durkheim's *The Rules of the Sociological Method* (1964 [1895]), which he operationalized subsequently in his influential study of *Suicide* (1952 [1897]). In so doing, classical sociology sought to make a radical distinction between 'culture', the realm of the social, and 'nature'. The social realm was deemed to be the proper domain of sociology, distinct from and not reducible to the inert objects of the natural sciences, such as biology. '*Homo Sociologicus*', as David Morgan and Sue Scott (1993: 2) note, 'is clearly distinguished from alternative versions relying upon biological processes, instincts and all other forms of reductionism and essentialism'. Hence, the development of sociology was predicated on a separation and a privileging of the cultural realm, or more specifically, the social realm, over the natural. The radical cleavage between the natural sciences (*naturwissenschaft*) and the cultural sciences (*geisteswissenschaften*) in German sociology in particular, reinforced this separation of the social over the natural (Hughes 1974). In so doing, the body, viewed as a thing in nature or as matter, was given relatively short shrift. The spectre of biologism, then, in many ways, can be said to have stalked the sociological tradition (Morgan and Scott 1993; Williams and Bendelow 1998; Burkitt 1999) and contributed to the neglect of the body.

Moreover, the construct of the subject of sociology as a rational actor, and the privileging of the notion of rational action which, according to Weber (1976 [1905]), was deemed to be the cornerstone of modern industrial capitalism, 'meant that the body was not perceived as a source of personal knowledge or understanding, or deemed relevant to the production of sociological knowledge' (Howson and Inglis 2001: 299). Rational action, in effect, became the benchmark against which to measure all other forms of action, such as traditional or affectual action. These, in turn, were always found to be lacking and disorderly in comparison with the orderliness of rational action. The project of modernity, premised on enlightenment thinking, which linked the notions of reason and freedom to the growth of industrialism, science and technology, also sought to celebrate the triumph of culture over nature. It is interesting to note that it was not only the body that was neglected in classical sociology, other topics such as touch, emotions and sexuality, which are also associated in some way with the body, were also marginalized (Morgan and Scott 1993; Williams and Bendelow 1998); these, too, have become areas of increasing interest in contemporary sociological discourse.

The foundation of modern sociology was forged out of the privileging of the social elements over the biological or the psychological in the classical tradition (Hirst and Woolley 1982). In this way, sociology reinforced the mind/body, culture/nature dichotomies in western humanist thought. These binaries had consequences for the visibility of particular social groups in sociological discourse. Women, for example, who were increasingly biologized in nineteenth-century thinking through association with their reproductive capacities, and thus consigned to the nature side of the culture/nature dichotomy, remained effectively outside of sociology's gaze, which was directed towards the social, public (male) sphere (see Sydie 1987; Laqueur 1987). At the same time, the fact that the body was associated with nature and femininity 'further distanced it from sociological analysis' (Howson and Inglis 2001: 298).

The researcher, too, in scientific sociological accounts became disembodied through the dictates of observation and objectivity, which constituted the watchwords of the positivist tradition. Recall Durkheim's first and 'most fundament rule' of sociological method: 'Consider social facts as things' (1964: 14). The first corollary to the principle rule is '*All preconceptions must be eradicated*', which, Durkheim argues, is 'the basis of all scientific method' (31). In other words, in order to get to the truth you have to remove yourself from the social world you inhabit and observe it at a distance through a disembodied gaze. Ironically, in distancing itself from 'nature', sociology has often taken the body as a biological given rather than questioning naturalist explanations of the body (Benton 1991). Thus, rather like societal members, it may be suggested, sociology has tended to view the body as an unquestioned 'natural' fact of existence, while offering disembodied accounts of social action. Morgan and Scott (1993: 5) argue that this unquestioning attitude towards biological explanations of the body in sociology in the past opened the way for 'the increasing influence of sociobiology, which takes ideological understandings of the natural inevitability of certain bodily processes and practices and presents them back to us cocooned in scientific language' (see also Hirst and Woolley 1982 on this).

Given these various reasons as to why sociology did not treat the body as a serious topic of inquiry, and there are others, too (see B. Turner 1991; Frank 1991), how is it that we have been in the midst of a kind of 'body fever', not only in sociology but, as indicated above, across the range of social sciences and the humanities in recent times? One simple answer to this is that the body was not quite as absent as was formerly thought in academe. Turner (B. Turner 1991), for example, makes a strong case for arguing that while the emphasis of culture over nature in classical sociology meant that sociology did not consider the ontological status of human beings, anthropology, at least in its early phases, did ask questions about human nature. As a consequence, while the body lay quiet but nevertheless implied in sociology, it became a topic of anthropological enquiry. As Morgan and Scott (1993) suggest, the relative absence of the body in sociology in comparison with anthropology could have been a further example of sociology attempting to establish its independence from a rival discipline. Chris Shilling (1993) argues that although the body was not central to sociology, nevertheless it had a secret and furtive presence in the discipline. Others, such as Simon Williams and Gillian Bendelow (1998) argue that 'the body' in classical sociology is recoverable by re-reading the classic texts in light of theoretical (corporeal) insights gained through the recent turn to the body in sociology.

It does not take too much imagination to re-vision *The Protestant Ethic and the Spirit of Capitalism* (1976 [1905]), in terms of a caged, rational, ascetic body, for example. Durkheim's later work on collective representations in *The Elementary Forms of Religious Life* (1968 [1912]) and his analysis of the oppositional pull between the sacred and the profane in primitive religions, led the way for others such as Marcel Mauss (1973 [1935]) and Mary Douglas (1970, 1973) to develop his ideas in terms of body symbolism. Similarly, the centrality of 'creative labour' in Marx's (1975 [1844]) early writings leads one directly to the body. However, it has to be recognized that Marx, Weber and Durkheim were more interested in analysing the causes and the consequences of the emergence of industrial capitalism than examining the sensual bodily aspects of human existence. The reasons why the topic of the body has turned into a minor academic industry in recent years are more complex than this.

THE BODY PROBLEM IN CONTEMPORARY SOCIOLOGY

It has been argued that post-positivist developments in social and cultural theory (feminism, postfeminism, psychoanalysis, poststructuralism, postmodernism and social philosophy) have had a considerable impact in moving 'the body' to the centre stage of the social sciences (Frank 1990; B. Turner 1991; Shilling 1993; Davis 1997; Howson and Inglis 2001). These theoretical interventions, as will be discussed over the course of the book, precipitated a questioning of the traditional (positivist) approaches upon which sociology had been founded and developed, such as the perceived radical disjuncture between the concepts of culture and nature, logos and eros, mind and body, male and female, which, it was argued, were central to sociological reasoning (Howson and Inglis 2001). These various 'isms' sought to challenge the dualistic thinking upon which the western humanist tradition was predicated. At this point, I will offer a few significant examples, focusing on feminist, poststructuralist and postmodernist viewpoints.

Just as there are distinctions between different feminist viewpoints, there are significant differences between poststructuralist and postmodernist approaches as well as overlaps, although the terms are often treated as synonymous. However, although the differences can often be quite nuanced, for the purposes of this discussion, I will not make a clear distinction between them, although I have done so elsewhere (Thomas 1995, 1996; Thomas and Walsh 1998). Margrit Shildrick (1997) proposes that poststructuralism and postmodernism can be treated as synonymous in as far as they both constitute a critique of modernity and the liberal humanist tradition in particular. Poststructuralism, Shildrick suggests, in the strictest sense, encapsulates a range of theories that largely relate to the 'linguistic and philosophic context', and 'are characterised by a rigorous anti-humanism and anti-subjectivity, and by the claim that all knowledge is discursively constructed' (5). Postmodernism is evidenced in a wider range of contexts and 'has come to denote a range of cultural theories and practices which break with the unity and certainty of the Western intellectual tradition' (ibid.). Michel Foucault, for example, is sometimes characterized as poststructuralist because of his anti-essentialist, anti-humanist viewpoint and his emphasis on discursive practices, but he has also been labelled as postmodernist (Frank 1991) and he has had a profound influence on, for example, postmodernist feminism (see Nicholson 1990).

Arthur Frank (1990) rightly considers that feminist politics was an early important impetus for the recent turn to the body in social and cultural thought, and most body studies theorists, whether they draw on feminist viewpoints or not, pay lip service to this view (see, also Bordo 1993). The body was a key site for second-wave feminists, as it is now for postfeminist writers (influenced by postmodernist and poststructuralist thought). From an array of theoretical perspectives which developed from the 1970s, feminists have systematically challenged the ways in which women's bodies have been subjected and controlled by 'men, the state and state agencies' by documenting the exploitation of women's bodies 'via domestic violence, rape and sexual abuse, advertising and pornography, medical interventions, exploitation and harassment at work', and so on (Morgan and Scott 1993: 10). In order to undercut the 'biology as destiny' tag, which had all too often been drawn upon to explain women's lack of fitness for office and the public sphere, and to shore up gender inequalities, socialist feminists argued for a line to be drawn between the categories of sex and gender (see Oakley 1974; Sydie 1987). Sex was taken to refer to biological differences between men and women and gender to socio-cultural (man-made) differences. It was the cultural construction of gender differences, which became the central concern of much feminist theorizing from the 1970s to the mid-1980s. Other feminists (existential, liberal and radical) argued that women were in fact exploited as a consequence of their reproductive capacities and if they were to achieve equality, they would have to transcend their biology, by means of a range of strategies, such as the politics of separatism (Millett 1977), or by technological interventions which would free women from the burden of reproduction (Firestone 1970).

Although feminism sought to challenge the received dualisms of nature/culture and male/female etc., the tacit acceptance of biological explanations of differences between men and women and the privileging of the social construction of gender, meant that the dualisms remained intact, as more recent feminist theory, informed by poststructuralist and postmodernist thought, has argued (see Gatens 1996). Nevertheless, as Frank (1991: 42) notes, 'feminism has taught us' that the story of society, on which many of the questions of power and control rest, 'both begins and ends with the body'.

The impact of Foucault's (1973, 1977 and 1984) work on disciplinary technologies and the technologies of the self has been of particular importance across the social sciences and the humanities in directing attention to the body. His anti-humanist stance rejects the idea of the free-willed, conscious individual as the cornerstone of the history and supplants it with a consideration of how bodies and subjects are moulded and fashioned through a variety of discursive practices that legitimate particular forms of domination. The body is the starting point for Foucault's (1986) notion of 'effective history'. The body, conceived of as an unstable, unfinished entity, becomes the key site for the operation of modern regimes of power, which, in contrast to earlier regimes, are not repressive but rather are 'productive'. Foucault's anti-essentialist, anti-humanist analysis of bodily surveillance in his studies on the prison, the asylum and the clinic, along with the history of sexuality, has had a profound impact on 'understanding the body as an object of processes of discipline and normalization' (Davis 1997: 3). Following on from this, the radical discursive approach of feminist philosopher Judith Butler (1990, 1993) has significantly impacted on the field of body studies, feminist and queer theory by challenging unquestioned 'naturalist' notions of sexual difference, which, as suggested above, ironically pervaded much of earlier second-wave feminist scholarship. This will be discussed more fully in Chapter 3.

In her keynote speech at the 'Body in Society and Culture' conference in 1990, Emily Martin, the feminist anthropologist, proposed that the then burgeoning fascination with the body, in part, may have been due to the fact that the body has a central place in western social formations. She also suggested that it may have also been related to the fact that 'we are undergoing fundamental changes in how our bodies are organized and experienced' (Martin cited in Csordas 1994: 1). Further, she proposed that we were in the process of experiencing 'the end of one kind of body and the beginning of another' (ibid.).

Contemporary scholarship in the social sciences and the humanities, according to Thomas Csordas (ibid.), appears to support Martin's contention. The classic notion of the 'old body' in both popular and academic thought, as indicated above, is a fixed bounded entity governed by the rules of biological science, which exists prior to the 'mutability of cultural change and diversity' and is typified 'by unchangeable inner necessities' (ibid.). The 'new body', on the other hand, 'can no longer be considered a brute fact of nature' (ibid.). Rather than taking the body for granted as an alterable fact of existence, as it had been in positivistic approaches to sociology, for example, its very status has been problematical. Hence, to a certain degree, this new body confronts the old body, which, I have suggested, most people tend to hold in their heads as they go about their everyday life. Following Foucault, it is now proposed that the body has a history; it should not be viewed as a fixed entity in the midst of (social) change. Rather, for some, it should be regarded as the very essence of change.

Frank (1990, 1991), for example, suggests that the resurgence of interest in the body as an object of theoretical discussion is not only bound up with feminist interventions but also with tensions and contradictions in modernist discourse, which were highlighted by and to a certain extent carried forward by, postmodernist discourse. In modernist discourse 'the body is the only constant in a rapidly changing world' (Frank 1990: 133). However, he notes that this positivist notion of the 'natural body' as a constant was challenged by social commentators who argued for a social constructionist perspective on the body on the evidential basis of the diverse ways in which the body is viewed and used in different cultures (see Polhemus 1975, 1978; Frank 1990, 1991). Both these positions, according to Frank (1991) are to be found in modernist discourse. Frank's view of the twofold 'modernist impulse' in social thought is reminiscent of Charles Baudelaire's definition of modernity, which he characterized in 1861 as, 'the eternal and the immutable' on the one hand, and 'the transient, the fleeting, the contingent' on the other (Baudelaire cited in Harvey 1989: 10).

Postmodernism, as the word suggests, indicates a shift or break with modernism. To put it in simple terms, postmodernism implies a rejection of universal truth and knowledge claims in favour of fragmentation and 'dispersed competing perspectives' (Shildrick 1997: 6); the abandonment of the 'grand narratives' of modernist (liberal humanist) thought founded on a universal construct of rationality; the questioning of the validity of the notion of human subjectivity and individual claims to knowledge; the erosion and blurring of the boundaries of discrete, hierarchical modes of knowledge, particularly the division between high and popular culture, and theory and practice. As such, the body constitutes a key knowledge base to launch a critique of the 'grand narratives' of modernist discourse, which, founded on enlightenment philosophy, have tended to privilege 'the experience of the disembodied, masculine Western elite' (Davis 1997: 4). By taking up an embodying theory approach, the grand narratives can be deconstructed to reveal that they are just one narrative among many, not the absolute, fixed truth. The 'high-theoretical postmodernism of the body theories', according to

Frank (1991: 40), is discernible in the writing of a number of thinkers such as, Roland Barthes (1985), Jacques Lacan (1977), Giles Deleuze and Felix Guattari (1983), and Foucault (1977, 1980, 1984). Frank, however, maintains that postmodernism not only breaks with and challenges modernist conceptions of the body, but it also carries forward the tensions and contradictions that are evident in modernist approaches. For example, he suggests that the body in postmodern discourse loses its material basis and becomes a metaphor. On the other hand, postmodernists who have followed in Foucault's wake, have used the body as the key site to examine how different subjectivities are constructed or to explore the multiple operations of disciplinary power. Thus, as Kathy Davis (1997: 4) has argued, 'both modernist and postmodernist scholars alternately propose the body as secure ground for claims of morality, knowledge and truth *and* as undeniable proof for the validity of radical reconstructionism'. In light of the above discussion, one explanation as to why the body moved up the sociological agenda concerns the development of new conceptual outlooks and tools, along with theoretical controversies and contradictions within these.

As Martin's words cited above suggest, shifts in academic interest and discourse do not occur in a vacuum. Sociologists are touched by societal changes that impact on and interact with, everyday life and they are often drawn to explore and engage with these through their research. It has been argued, for example, that the ascent of the body in sociology speaks to and of a variety of interrelated shifts in late twentieth-century capitalism through which the body was marked as a key site: the increasing penetration of consumer culture into almost every aspect of our bodily existence (Featherstone 1991; Falk 1994); the shifting politics of identity and difference and the loss of a stable, fixed sense of self (Giddens 1991; Martin 1994; Butler 1990, 1993); accelerated technological developments in the field of bio-medicine which, for example, can capture, reveal and repair parts of the body that hitherto could not be seen by the naked eye (Martin 1994; Thrift 2004); demographic shifts in western cultures towards an increasing ageing population (Featherstone and Hepworth 1995); the risks and moral panics associated with the advent and global reach of HIV and AIDS (Kroker and Kroker 1987; Shilling 1993; Shildrick 1997; Phelan 1997); the risks associated with the environment, pollution and wars, and since '9/11' and '7/7', fear of 'global terror' and human bombs. It seems, then, that we have entered into what Turner (B. 1992) calls a 'somatic society'.

Because social analysis is by nature critical, recent developments in and approaches to, the study of the body have also been subject to debate and critique. A common criticism of the bulk of the work in this field is that it is saturated with theory at the expense of detailed empirical analysis (see for example, Howson and Englis 2001; Shilling 2001; Crossley 2001). This imbalance, however is beginning to be addressed more in certain studies (see, for example, Wacquant 1995, 1998; Davis 1997; Nast and Pile 1998; Evans and Lee 2002; Thomas and Ahmed 2004; Crossley 2006; Shilling 2007). Another is that 'body studies' tend to use the body as a focus for studying something else, such as consumer culture, gender, race and ethnicity, risk, health and illness, technologies of the body. As a consequence of the dominance of representational or discursive models, much influenced by the work of Foucault, it is argued that 'the body' simply disappears as it is brought into the analytic frame of reference (Shilling 1993). In spite of challenging the mind/body dualisms inherent in the dominant tradition of western humanist thought, it is argued that the majority of the work on the body unwittingly maintains and supports them by privileging a social constructionist position (B. Turner 1992; Shilling 1993; Williams and Bendelow 1998; Burkitt 1999). By focusing on representation, according to corporeal feminists (Grosz 1993, 1994, 1995) and phenomenologically oriented thinkers (Csordas 1993, 1994, 2002) for example, body studies overlook the importance of the situated experience of being a body in society and the fact that the body is an unstable, unfinished entity. The domination of theories of representation, it is argued, need to be counter-balanced by a thoroughgoing exploration of the 'lived, experiential body' (Csordas, 1993, 1994, 2002; Crossley 2001b), which often ends up drawing on Maurice Merleau-Ponty's (1962, 1968) phenomenological approach to embodiment. Some writers suggest that this could also provide a basis for exploring the relations between the social and the natural body (B. Turner, 1992) in a way that overcomes the nature/culture divide, which has characterized the sociology.

Pierre Bourdieu's sociological framework has had a considerable impact on the development of cultural sociology and body studies, as a quick scan of the back issues of influential journals like *Theory, Culture and Society* and *Body and Society* amply demonstrates. Bourdieu's theoretical frame, developed over many years, was shaped by elements of his training in philosophy, particularly aspects of phenomenology, anthropology and sociology, and Durkheim especially (see Bourdieu 1990a;

Bourdieu and Wacquant 1992 for in-depth discussions of his intellectual influences). Like Foucault, Bourdieu views the body as an unfinished entity, but he is opposed to a radical social constructionist approach, of which Foucault has often been accused. Bourdieu seeks to overcome what he sees as the objectivist/subjectivist dualisms or 'deterministic scientism and subjectivism or spontaneism' (Bourdieu 1993: 55) as he calls it, which he considers have haunted sociology. The former, for Bourdieu, is typified by over-determining structuralist approaches which remove the 'social agent' from the analytic frame. The latter is represented by writers who focus on a subjective, existential, experiential approach to the exclusion of social structure.

The body, for Bourdieu (1984, 1990b) is a carrier of symbolic value, which has become a form of commodified 'physical capital' in contemporary society and which is 'produced' by acts of labour. These acts of labour affect the way in which people, or 'agents' in Bourdieu's terminology, develop and preserve their bodies and the ways in which they represent themselves on a daily basis through body techniques, dress and style, for example, which, for Bourdieu (1984), inevitably carry the indicators of the individuals' social class. Bourdieu's conceptual framework of 'field', 'capital' ('physical', 'social' and 'cultural' too) and 'habitus', which seeks to address the complex relations between 'objective structures' and the multiple ways in which the bodies of social agents are 'made', has had an impact on research on physical practices such as boxing (Wacquant 2004) and dancing (Wulff 2003; Wainwright *et al.* 2005; Morris 2006) and the sociology of sport generally, although he also has detractors in this area (see Khon 2003).

The idea of what constitutes 'the natural' has also recently been reconsidered through the impact of concepts such as 'performativity' (Butler, 1990, 1993), 'cyborg' (Haraway 1991) and 'posthuman'. In different ways, these relatively new, influential constructs have respectively challenged the culture/nature and human/non-human dichotomies, which the western tradition of social thought has been so eager to uphold. Needless to say, to what extent they are successful in so doing is another matter and one which is explored in different parts of this book.

CONCLUSION

Despite the criticisms outlined above, the now not so new social study of the body has raised important theoretical and methodological questions regarding traditional social and cultural analysis. It has also pointed the way forward for social and cultural studies in the early part of the twenty-first century by promoting corporeal theories that highlight the fluid and shifting, yet situated character of the body in society, which, in turn, have implications for social relations in an era of new technologies and global market economies. This book aims to draw out and examine selected key concepts, methods and critical insights gained from the study of the body. In so doing, it seeks to consider the implications for the development of future socio-cultural analysis.

As indicated earlier in the chapter, rather than covering the full spectrum of topics associated with the social study of the body, this book uses a case study approach which draws on certain aspects of performance practice: dance and music, dress, photography, boxing and live art, to set out and make sense of contemporary approaches to the body. These practices may not be viewed as 'everyday life' in the usual sense of the term. However, they often draw on the everyday or emerge out of the vernacular. When American modern dance emerged in the second decade of the twentieth century, modernist iconoclasts such as Martha Graham, for example, rather like Durkheim in sociology, sought to eradicate all preconceptions concerning what dance is and return to 'first principles'. Graham turned towards the impulses of everyday movement and gesture in order to create choreographies based on what she called the 'stuff' of dance, which was significant (meaningful) movement (Thomas 1995).

Graham's experimental dance style was formalized into a codified technique over the years, so that the everydayness of gesture was lost. In the 1960s, postmodern dancers, who were reacting against the codification of expressionist modern dance and its star system, took everyday gesture as action in and of itself, not as a referent for something else like an emotion or feeling, but as both the starting and end point of their dances. Moreover, practices such as music (see Bull and Back 2003) and dance are central to the lives of many individuals and groups, as I will demonstrate in the discussions of images of ageing. It might be suggested that these so-called 'aesthetic' practices are in conflict with the demands of social science, where the emphasis is placed on 'science'. Sociologists, after all, have traditionally bracketed out the 'aesthetic' when studying the social significance of art (see, for example, Weber 1978 [1921] and Bourdieu 1996). With the breakdown of disciplinary boundaries and the 'aestheticisation of everyday life' (Featherstone

1991), the aesthetic is no longer necessarily out of sociological bounds and it could be argued, may be necessary to it (Inglis 2005; Wolff 2008). Indeed, the melding of performance and research, as in 'performance ethnography' (Mienczakowski 2001), following on from the work of anthropologists like Victor Turner (1986), has attracted the attention of leading qualitative sociologists (see, Denzin 1997).

This chapter briefly set out the groundwork for exploring the exponential rise of interest in the body in social and cultural analyses since the 1980s, and the key issues and concerns that confront the development of this work. The following four chapters draw on a range of performance practices where the body in some guise or another is centre stage, in order to think through some of the theoretical and methodological issues that confront sociological and cultural analyses. The substantial 'performance' case studies, however, are to be found in Chapters 3, 4 and 5. The choice of case studies was not entirely arbitrary; rather they seemed to fit well and exemplify the key concerns explored in these chapters. At the same time, they were by no means the only ones that could have been selected. The case studies were intended initially to stand alone in the context of the chapters. However, as the writing progressed it became evident that they also built on each other to some extent and that those interconnections between the themes of gender, ethnography and performance ran through the chapters in the context of thinking through the key issues around the body which lie at the core of the book.

Chapter 2 explores the uses of the constructs of 'performance' and 'performativity' and related terms, which have had an impact on the contemporary social and cultural study of the body. It does this from the vantage points of several subject areas, such as performance art, anthropology, sociology and linguistics. It considers a range of practices under the headings of aesthetic or cultural and social performances, in which ideas concerning the body are closely implicated and entwined. The ideas that emerge from these theories and practices are not clear-cut. Rather, they often flow into and inform each other, which also means that the distinctions between aesthetic and the everyday bodily practices are not necessarily as straightforward as may be imagined.

In Chapter 3, the 'body' in question is mostly centred on shifting ideas surrounding women's relations to their bodies in feminist discourse, which, as indicated above, has contributed a great deal to the interest and development of the study of the body in social and cultural analyses.

The changing themes within feminist discourse, from gender feminism to postmodern feminism, for example, are explored and subsequently raised in relation to plastic surgery, which has become a topic of feminist debate in recent years. The discussion is woven around two performance case studies, which I suggest speak to two different moments in feminist discourse; the first is the photographer, Jo Spence, and the second is the performance artist, Orlan. In particular, I focus on their respective performative practices which confront the subject of the (female) body and surgery.

In Chapter 4, the 'body' focus is considered in relation to ethnographic research, particularly in sociology and also anthropology to reveal the ways in which the matter of bodies is strewn across the ethnographic landscape, although not always intentionally. It considers the complex near/far relations between the researcher, the researched and the (sociological/ethnographic) field. A range of ethnographic devices, histories and practices are explored in which the researcher's bodily presence, the participants and the embodied field are interleaved, although these often go unnoticed or unreported. The principle tropes of ethnography are explored further through an ethnographic case study on boxing, Loïc Wacquant's (2004) fascinating study of his lived experiences of becoming an apprentice boxer in a Chicago gym, which is situated in a ghetto area of the city.

In Chapter 5, the focus on the 'body' centres on the ways in which older bodies have been constructed and perceived. It describes briefly the main approaches to ageing and society and the shifting paradigms and concerns that have dominated over the years. In so doing, it addresses the growing interest in the ageing body, in the context of the social and cultural factors that have contributed to this development. The chapter draws on performance case studies, including research on social and professional dance to explore the relation between the competing paradigms on the ageing body in contemporary society.

2

PERFORMING THE BODY

INTRODUCTION: PERFORMANCE AND PERFORMATIVITY

The focus of this chapter is directed towards 'uses' of 'performance' and 'performativity' in a range of aesthetic, cultural and social discourses. Just as the study of the body has gained currency in contemporary social and cultural analysis, so there has been a renewed interest in focusing on notions of performance and 'performativity', through the influence of thinkers such as Jacques Derrida (2001 [1978]) and Judith Butler (1990, 1993) in particular, whose 'postmodern performance theory' (Morris 1996) has made a significant intervention to the radical re-thinking of sex and gender and with this, notions of identity and subjectivity. These renewed approaches have also impacted on literary theory and theatre and performance studies. The idea of performance and performativity, however, has a longer history than this and is scattered over a variety of discourses, which do not necessarily correlate, although there are a number of overlaps and borrowings of theories and themes. In many ways, the notion of performance can be just as confusing and problematic as that of postmodernism, with which it is mostly associated (ibid.).

Approaches to performance in social and cultural studies have generally drawn sustenance from the practices and conventions of (mostly western) theatrical performance traditions and in a number of instances, have been criticized for so doing, particularly when discussing 'other' cultures' traditions and practices. The most important reason for focusing on different uses of performance in this chapter rests on the fact that ideas about the body, self and identity are heavily implicated in the very idea of performance, both older and newer approaches. A focus on performance will bring to light certain key concerns of and problems with the contemporary study of the body, which I will seek to explain, explore and develop in this and subsequent chapters. Moreover, as indicated above, the complexity of meanings and uses of the terms performativity and performance can be aligned with that of post-modernism and its relation to modernism, as well as forming a nexus between feminism and postmodernism (Carlson 2004). As such, it is important to consider the ways in which the terms performance and performativity have been used in the social sciences,. and anthropology and sociology in particular.[1] In so doing, the discussion will also touch on the development of performance studies, which initially drew heavily on social and anthropological notions of performance.[2] My main interest, however, as indicated in the previous chapter, is directed towards social and cultural analysis and thus, the way in which I draw selectively from other fields needs to be considered in this light.

The idea of performance has generally been used in two ways in broadly based social science; the first usage refers in the main to aesthetic or cultural performances such as ritual, theatre and dance, and the second is associated with performance practices in everyday life (Schieffelin 1998). Bringing these two areas together in one rather large chapter may seem rather unwieldy. However, it will become evident in due course that it is sometimes difficult, if not impossible, to separate the aesthetic from everyday notions of performance and performativity. A lengthy chapter discussion will add to the continuity of ideas that dart back and forth across this tenuous analytic divide. To set the scene, the chapter begins with a brief and necessarily general discussion of a particular (although varied) kind of critical artistic practice that became known as 'performance art', which also brought the body into critical focus.

PERFORMANCE ART: ACTIVATING THE BODY

The term performance, as Michal McCall (2000: 421) points out, 'entered art critical and academic discourses in the 1970s, to name a new visual art form and to distinguish dramatic scripts from particular productions of them – that is, from performances on stage'. This emergent form pitted itself against the normative conventions of traditional theatre and concert performances and in so doing, placed an emphasis on the 'body and movement' (Carlson 2004: 128). The theatre, in effect, became performance art's 'other' against which it sought to distinguish itself, whether or not the performance artists themselves had a theatrical background. Performance art such as body painting and radical theatre, involving an 'interdisciplinary and often multi-media style production' (Sayre cited in Denzin 1997: 107), had its antecedents in the *events and happenings* of the 1950s and 1960s. For Roselee Goldberg (1979), as for a number of writers, the roots of performance art lie much deeper and are to be found in the earlier twentieth-century avant-garde movements of dadaism, futurism and surrealism. However, Marvin Carlson takes issue with this direct-line approach from contemporary performance to the historical avant-garde tradition, which he argues 'can limit understanding of both the social functioning of such art today, and also how it relates to other performance activity in the past' (2004: 85).

The emergence of 'live' performance art in the US, as sociologist Norman Denzin (1997: 124) points out, 'corresponds to the politicization of art and culture associated with the Vietnam War and Watergate, as well as the women's movement'. This was an era in which a burgeoning second wave feminist movement was challenging traditional conceptions of femininity and seeking to wrest control of women's bodies from a male-dominated medical discourse. This challenge was exemplified by such blockbuster female body reclamation manuals as *Our Bodies Ourselves* (The Boston Women's Health Book Collective, 1973), which its sub-title declared was 'a book *for* and *by* women' (my emphasis). Feminist discourse clearly had an impact on women performers in the 1970s (Goldberg 1979; Harris 1999; Carlson 2004). But there is not just a one-way route from theory to practice, rather I would want to argue that artistic practice can provide a challenge to theory through performance and anticipate shifts in theoretical discourse. Take, for example, the dance and subsequent film work of Yvonne Rainer. Rainer was a dance iconoclast in the 1960s and early 1970s before leaving her performance career behind to become a full-time film-maker in 1973 (she had been using film in her dance work before this time). Her most celebrated dance work, *Trio A* (1966), which was designed to be danced by virtually anyone, constituted 'a signal work for Rainer and for the entire post-modern dance movement' (Banes 1980: 42; see also Wood 2006). Rainer's minimalist movement vocabulary, which involved mundane, workaday movement as a key component, went hand-in-hand with her polemical anti-dance establishment 'aesthetics of denial' manifesto targeted at the specular, virtuosic, transcendent, star imaging inherent in both classical ballet and modern dance performance systems. As a consequence of these combined factors, it was Rainer, much to her chagrin, who was more often than not identified as the leader of the emergent postmodern dance movement in the US.

In western theatre dance, the dancer's body is generally the main medium of expression, presentation and representation and thus the performer and the performance *seem* to be as one. In *Trio A*, Rainer sought to challenge what Peggy Phelan (1999: 6) terms the 'polarized operations of narcissism and voyeurism integral to dance spectacle' by 'disallowing the customary exchange of gaze between the dancer and the audience'. Rainer, Phelan argues, only succeeded in displacing the voyeuristic nature of dance spectacle rather than eliminating it. Voyeurism entails looking at someone without being seen, as in film for example, where the viewer/voyeur remains formally free to gaze at the image of the performer on the screen without fear of being looked at in return. In dance, however, there is always the possibility that the spectator will be seen by the performer; the dancer generally projects the body and the gaze outwards towards the audience with the intention to be looked at in return. By averting the dancer's gaze away from the audience in *Trio A*, Rainer 'produced a proto-cinematic relationship between spectator and performance' (7). That is, the dancer invoked what came to be known as the 'to-be-looked-at-ness' (Mulvey 1975) of the female protagonist on film. Rainer's concern to undermine the 'duel operations of narcissism and voyeurism' in dance performance in *Trio A*, according to Phelan anticipated later feminist concerns with the 'male gaze': '*Trio A* in particular anticipates feminist film theory's attention to the structure of the gaze in terms that are resonant with Laura Mulvey's celebrated 1975 essay "Visual Pleasure and narrative Cinema" ' (1999: 8).

Mulvey's (1975) essay on the unequal structure of 'looking' in narrative cinema, which was underscored by a psychoanalytic semiotics much influenced by the work of Jacques Lacan (1977), sparked off a minor gaze theory industry in feminist film theory. It also spread to other areas of cultural analysis, such as cultural studies and a burgeoning feminist dance criticism in the 1980s (Thomas 1996). Mulvey's approach was also subject to criticism (Gamman and Marshment 1988) and to revision (de Lauretis 1986; Doane 1987). The import of Mulvey's analysis for feminism, as I have shown elsewhere (Thomas 1996), lay in the fact that it offered a theoretical frame for linking the reification of women's bodies in the domain of cultural representation to their unequal access to cultural power which was embedded in the discourses of patriarchy. However, the theory was criticized for positing a universalist, static, undifferentiated, all male, heterosexual gaze, although Mulvey (1989) tried to remedy some of these problems in a re-worked version some years later.

As Phelan points out, the displaced dancer's gaze in Rainer's *Trio A* also anticipated a more radical and critical intervention into the presentation/representation problematic which is evident in her 1985 film, *The Man who Envied Women*, in which the female protagonist, Trisha, remains sight unseen throughout the duration of the film, with the sound of her narrative voice marking her absent-presence. Like the spectator, Trisha's bodily image is off screen, 'elsewhere' or 'space off' (de Lauretis 1986: 26); her body is inferred by the space taken up by her ever-present narrative voice in the film. *The Man who Envied Women* disrupts or fractures the filmic convention of the necessity to image the female protagonist and swathes the spectator in the 'elsewhere'. By posing the question of what happens when the object of the gaze is not represented by her bodily image, Rainer not only takes Mulvey's theory to its logical end, she also points to its limitations and challenges its legacy. Moreover, according to Phelan (1999: 8), she does so from the performance viewpoint, which 'both defines and shapes her filmic practice':

> Rainer saw very early on that the performer plays an integral part in representational economies which capitalize on 'the narcissistic-voyeuristic duality of doer and looker'. . . . As both a critique of the limitations of Mulvey's thesis . . . and a powerful illustration of the consequences of it, Rainer's post-1975 films and essays advance feminist film theory by challenging the visual/literal consequences of it – primarily from the perspective of performance.
>
> (ibid.)

Rainer's challenge to the apparatus of film and film theory, given impetus by her performance background, occurred at around the same time as the French critique of 'ocularcentrism' (Jay 1999) began to raise its head in the US. Ocularcentrism is a critical construct which refers to the dominance of the visual in the tradition of western culture. It was advanced by anti-humanist thinkers such as Derrida, Michel Foucault, Jean-François Lyotard and Julia Kristeva, and had a considerable impact on American art criticism in the 1980s (see Krauss 1985; Huyssen 1986).

Feminist dance scholars and practitioners such as Anne Daly (1987, 1987/88) and Cynthia Novack (1993) critiqued the primacy of the visual in ballet and the ways in which it objectified the female dancing body. The critique of visualism which ocularcentrism entails also had an impact on more traditional disciplinary fields, such as anthropology. The influential postmodern collection edited by James Clifford and George Marcus, *Writing Culture*, published in 1986, which spearheaded what came to be called the 'new ethnography', is a striking example of this critique.

Cultural commentators have also argued that women's performance has played a pivotal role in the nexus of feminism and postmodernism. Jeannie Forte (cited in Harris 1999: 28), for example, suggests that while all performance art, at least in its early stages, 'evidenced a deconstructive intent', women's performance in particular 'emerges as a specific strategy that allies postmodernism with feminism'. Geraldine Harris also notes that: 'the work of American female performance artists of the 1970s and 1980s and its reception by feminist critics and theorists have had a notable influence on the subject area in the USA' (ibid.).

Performance art has gone through a number of shifts and transformations from the 1970s to the present day, which resonate with the turn towards and the subsequent unfixing of the (gendered) body in culture as a discrete entity in social and cultural analyses, given impetus by (post-structualist) ideas that the world we inhabit is one of flux, instability and diversity. In turn, this culminated in recent theorizations of the post-human, cyborgian and virtual body, which broadly speaking, addresses the relations between the body, animals and machines within the context of technology and culture (Murphie and Potts 2003).

The following quotation from Denzin (paraphrasing and citing the performance activist/theorist, Johannes Birringer (1993: 121–2)) provides a concise summary of those transformations in performance art:

> Transformations of this [performance art] movement focused on the gendered body, dissolving fragmenting and displacing its centred presence in performance. . . . The literal body was challenged, replaced by a gendered, autobiographical, confessional body text, cries and whispers on stage, gyrations and repetitions, and a body in motion working against itself and its culture. . . . This body was soon replaced by a globalizing electronic, postmodern video technology. Laurie Anderson's multitrack, audiovisual choreographies are key here. Her electronic body represents a kind of

> 'closed circuit transvestism' . . . awaiting only on-line movement into cyberspace . . .
>
> (Denzin 1997: 107)

In a wide-ranging overview of approaches to performance, Carlson (2004: 128) argues that 'the initial oppositional stance of performance art to theatre in the 1970s and 1980s has been steadily eroded', while the 'emphasis on body and movement' over language in 'performance activities', has gradually given way 'to image-centred performance and a return to language'. Carlson sees that the reinstatement of verbal discourse, even in the work of the most physically oriented performance artists, reflects artistic, social and political concerns to give 'voice' to individuals and groups who are excluded from or who have no active role in the current system: 'This sense of providing a voice and a body to common (and generally unarticulated) experience is very important to much modern performance, especially that created by and for marginalized or oppressed communities' (ibid.).

Nevertheless, these observable shifts do not represent a return to the normative theatrical tradition of old. Rather, as Denzin (1997: 107) notes, in contemporary performance art 'the text becomes dialogical'. The tendency towards improvisation, as opposed to recitation of a preordained script, means that the 'authority' of the text is undermined. That is, the notion of the author, choreographer or performance artist as the sole creator of the authentic fixed 'work' is challenged and rendered invalid, heralding what Barthes referred to in relation to literature as 'the death of the author' (1977: 142–8). As a consequence of this dialogic move, 'the emphasis is on the performance' (action, process) in a dialogical engagement with the audience and 'not the work' as a thing in itself. As we shall see in the following chapter, the shift towards action and process as opposed to the (con)textual is also evidenced in the uses of performance and performativity in social and cultural studies. This is despite the fact that performance has been used in somewhat different ways in the social sciences, linguistics, theatre studies and theatre anthropology for a number of years prior to the 'crises of representation' (Boyne and Rattansi 1990) brought about by postmodernist sensibilities.

It is interesting to note that although the term 'performance art' waned in terms of its importance towards the dawning of the new millennium, the word performance continued to be important and indeed, developed in ways which were 'often far removed from the rather abstract art world origins of the movement' (Carlson 2004: 134). It is worth pointing out that at around the same time as performance art began to reincorporate verbal discourse into the work, social and cultural analysis, with the shift towards the 'cultural turn' in the sociology, became increasingly interested in directing the focus of attention towards the corporeal aspects of culture. At the same time, some sociologists and anthropologists concerned to get away from the objectification of culture, articulated through the authorial voice of the ethnographer, began to turn their field notes into performances, with a view to generating 'performance ethnography' and 'ethnodrama' approaches (see, V. Turner 1986; Ness 1996; Denzin 1997; McCall 2000; Saldana 2005, 2011).

There are significant differences between the two genres of performance ethnography and ethnodrama. The former may be seen as 'postmodern' in as much as it operates on the basis of 'blurred genres' (Geertz 1983) through employing strategies such as parody or irony (see Mulkay 1985; Richardson 1997). Performance ethnography occupies 'a space in ethnographic discourse which challenges traditional reporting approaches through the incorporation of genres, practices and techniques' employed in theatre, film, ethnography and performance (Mienczakowski 2001: 469). According to Jim Mienczakowski (ibid.), echoing Denzin's view (1997), the move towards ethnographic performance 'is a logical turn for a number of human disciplines in which *culture* is increasingly seen as performance and *performance texts* as being able to concretize experience'.

Exponents of ethnodrama, on the other hand, do not simply attempt to transgressively blur the boundaries between genres; rather they are concerned to generate 'a form of public voice ethnography that has emancipatory and educational potential' (ibid.). This means that the data-gathering techniques and the validation processes embedded within them require extensive consensual agreement between participants and researchers in order to generate more 'plausible accounts' (Atkinson 1990). These credible, agreed explanations are forged through the dialogical process, thereby potentially blurring the boundaries between professional and lay members' interpretations of reality. Although ethnodrama may sit within a 'school of theatre which searches for social change', it nevertheless diverges from theatre by insisting on the importance of the formal methodological principles and tools of ethnography over and above 'the artistic demands of aesthetics' (Mienczakowski

2001: 471). Thus, there remains a tension between the aesthetics of theatre and the theatre of everyday life, with verisimilitude taking precedence over aesthetics. The entry and/or intervention of performance into the hallowed ground of social science methods also brings the body into focus, as it is inevitably in performative action. As such, it might be expected that the physical, embodied presence of the ethnographer, who, after all, is enmeshed in the very processes of the research itself, is factored into the analysis. As Amanda Coffey (1999: 75) has noted however, ethnographers in general have paid scant attention to 'the embodied nature of fieldwork itself' from a methodological point of view. As will be discussed later in the chapter, ethnographers, like Kirsten Hastrup (1995), for example, argue for an embodied approach to social research which utilizes the tools of 'theatre anthropology' as a way forward.

USES OF PERFORMANCE: AESTHETIC, CULTURAL AND SOCIAL PERFORMANCES

In assessing the advantages and problems associated with employing terms like performance and performativity in anthropology, Edward Schieffelin (1998: 194) notes that there has been a shift in recent years away from studying 'cultural performances as systems of representations' and towards a consideration of the processes and practices of performance. As a consequence of this shift, Schieffelin states that: ' "Performance" deals with actions more than text: with habits of the body more than structures and symbols, with illocutionary rather than prepositional force, with the social construction of reality rather than its representation' (ibid.).

Once actions are given priority, it seems, the body also moves to centre stage. In a similar vein, Richard Bauman, the linguistic anthropologist and folklorist, proposes that performance is both 'a mode of communicative behaviour and a type of communicative event' (1992: 42). The most common use of performance in recent years, according to Bauman, refers to the actual execution of a particular action, as opposed to the potential for action, or indeed, an abstraction from the action itself. Hence, both writers agree that there has been a shift away from the performance-as-text model as exemplified in Clifford Geertz's (1975) interpretive account of a Balinese cockfight as a form of western tragedy, and towards the study of 'actual performances' (Hughes-Freeland 1998: 3).

According to Schieffelin (1998: 194) the notion of performance has been used principally in two ways in the social sciences: the first refers to 'symbolic or aesthetic activities such as ritual or theatrical and folk activities', while the second is linked with performative activities in everyday life. Both uses draw on the conceptual apparatus and terminology of the western theatrical tradition, although until recently, most writers have drawn a distinction between theatrical performances 'on' and 'off' the stage. I will take each of these in turn, noting overlaps and differences. The reason for so doing is to offer a more solid understanding of the complexity of terms like performance and performativity and how these impact on notions about the body in everyday life through the prism of extra-everyday life, for want of a better term.

AESTHETIC OR 'CULTURAL PERFORMANCES' AND THE EVERYDAY

In the first usage of performance, as indicated above, symbolic or aesthetic activities are viewed as 'intentional expressive productions in established local genres' (ibid.). The idea behind this is that tradition is carried through not only by people but by 'different forms of cultural media', which include but are by no means confined to, spoken language (Singer 1972: 76). In this view, cultural performances not only encompass traditional activities which come under the umbrella of aesthetic or cultural performances in the west, such as dance, theatre, concerts and lectures. They also incorporate those activities which are normally classified under 'religion and ritual' (71), such as 'prayers, ritual readings and recitations, rites and ceremonies, festivals' and so on.

Within this particular context, performance usually refers to special, marked out, bounded actions, which are separated off from everyday actions and which are intentionally produced and have some purpose or meaning for those involved in the event (performers and audience). Here it is employed to denote 'a display of expressive competence or virtuosity by one or more performers to an audience' (Schieffelin 1998: 195). It is viewed as a 'performative event', which is 'treated as an aesthetic whole in a larger social context' (ibid.). The focus of study in this approach is on the structure of these events, the way they are carried out, or the effect they have on the performers and audiences, and 'their relation to the social context' (ibid.). The event, if you like, seems to achieve a categorical 'framed' status that can be dissected to reveal something about the performance tradition and the culture under consideration.

The idea of the performance event, which led to the contextual approach in folklore research, contributed to the shift from the emphasis on 'text' to that of 'function'. Carlson (2004: 15) argues that despite the concern with the 'how' of performance, the performance event approach remains largely contextually bound as opposed to focusing on the 'actions' of the performer. The marking out or the framing of the performance as an event, assumes that it is this framing that allows for an audience or a culture to make sense of and experience 'the performance *as* performance' (ibid.). In this approach, cultural or aesthetic performances are viewed as something different from ordinary behaviour. For Bauman (1992: 46), for example, the importance of performances to the study of society 'lies in their nature as reflexive instruments of cultural expression'. The anthropologist Dell Hymes (1975: 13) argues that aesthetic performances are situated within the context of culture and that these have emergent properties which 'arise or unfold within that context'. In this sense, performance is not to be confused with ordinary (everyday) behaviour. It should not be perceived 'as something mechanical or inferior, as in some linguistic discussion', according to Hymes, but 'as something creative, realized, achieved, even transcendent of the ordinary course of events' (ibid.).

An examination of aesthetic bodily practices in the context of performance, it may be argued, can enable us to say something significant about the structures and processes that underpin taken for granted attitudes and modes of interpersonal communication in everyday life. Anya Peterson Royce's (1980: 27–31) observations on an erupting 'social drama', between particular members of a Zapotec dance group from Juchitán in Mexico during a group rehearsal, are instructive in this regard. This particular social drama arose over a heated dispute between two members of the group as to the correct way to perform the *fandango*. The *fandango* is 'an open couple dance that is regarded as one of the most traditional of the Zapotec dances' (28).

Once a year, 'the Zapotec of Juchitán send a delegation of dancers to the Guelaguetza in Oaxaca City, the state capital' (27), where, according to Royce, they perform alongside other groups of indigenous dancers and musicians from the state. In 1972, six men and six women dancers from Juchitán were chosen to represent the city at the annual festival. The selection process involved several factors. The women tended to come from older, richer families which were closely 'aligned with the political powers in the city' (28) and they were generally recognized as the best dancers. After the women were selected, the men were chosen on the basis of their 'friendship with the women dancers, their willingness to take part, and dancing ability' (ibid.). Royce does not comment on this but it appears that, as often happens in the UK social dance contexts, the Zapotec men are more reluctant to 'join the dance' than the women and they are perhaps less skilled, too, as a consequence.

The rehearsals commenced following the selection of the dancers. In the *fandango* 'the music and the dance alternate between fast and slow sections' (ibid.). At the beginning of each of the movement sections the man and the woman change places. During the rehearsal, Royce observed that two couples changed places at the same time as a change occurred in the music, while the other four couples began to change places two bars before the change in the music. An older, experienced female dancer, who was considered an expert on dancing and who had performed regularly at the annual dance festival, corrected one of the women belonging to the two couples who had moved on the change of tempo, who was also her younger cousin. The cousins belonged to a prestigious, old Zapotec family, the Gómez, as did one of the other women in the group, and a further two were connected with the family in some way.

Given the older dancer's position in the hierarchy of the Gómez family and the family's recognized authority on Zapotec dance tradition, the younger cousin would have been expected to accept the correction and come into line with the other four couples. However, she did not, maintaining that she and the other female dancer were correct in changing places with the change in the music and that indeed, she was only following the example of her corrector's grandmother, who was viewed as one of the very best dancers in the city. The other female dancer who also moved on the musical change belonged to a rival family, the Martínez, which, although very well connected, was relatively new to the city and therefore, its members were not perceived by the Gómez family to be an authority on Zapotec tradition. Despite appeals to traditional Gómez family loyalty and being reminded of the familial tradition of fine dancing, the younger Gómez dancer steadfastly refused to give way to her older cousin's viewpoint and the rehearsal came to an abrupt end, with the other three women connected to the dominant Gómez family taking sides with the older cousin against the younger cousin and the dancer from the Martínez family. The younger Gómez dancer insisted that her older cousin's grandmother should arbitrate on the matter.

When the case was presented to the grandmother by both sides, she asked both women to perform the *fandango*, after which she supported her granddaughter's version of changing places two measures before the change in the music. At this point, the younger dancer insisted that the older woman had herself crossed over on the change of the music at a recent festive family occasion. The older woman turned to her own daughter, also a recognized dancer in the family, and asked her if this was the case. The daughter sided with her mother (and by implication her own daughter, the older cousin), confirming that she had performed it on that occasion in the way the family had always performed it, moving two measures before the change in the music. The younger, recalcitrant Gómez dancer, as Royce notes, 'could no longer maintain her position in the face of the combined age, authority, and numerical advantage' (30) and had to give way.

However, some time afterwards Royce herself was performing the *fandango* with another woman from the Gómez family. As Royce moved to change places two measures before the musical change, she was informed that she should not do so until the music changed. After observing the dance numerous times and asking a lot of questions, it became clear to Royce that both versions were acceptable. Hence, she concluded that the 'cultural values upheld on the occasion of the rehearsal were those of Gómez superiority and necessary to show family allegiance and unity' (31). The confrontation over the correct dance movements in the rehearsal revealed a number of underlying cultural values and symbols embedded in Zapotec culture. The dancers' body movement in time and space in the context of the rehearsal became a site of resistance to and an affirmation of the cultural codes of behaviour which almost go unnoticed in everyday life. This case also raises the question as to when a performance event (in the case of a rehearsal) can be said to begin and end, which, in turn, leads to a questioning of the closed-off notion of the 'performance event' from everyday life.

The work of anthropologist Victor Turner has been of particular significance to the development of the analysis of cultural or aesthetic performances. Turner's approach to 'social drama' in anthropology, however, was initially conceived as a useful tool for analysing the organizational structure of a given (traditional) society, as opposed to the performative context, which writers such as Bauman and Hymes focused on (Carlson 2004). Hence, he drew on the 'structure of traditional dramatic action' rather than 'the performance situation of theatre' (16). Social dramas, according to Turner (V. Turner 1982: 69–71) have four phases, 'breach', 'crisis', 'redress' and either 'reintegration' or 'irreparable breach'. He does not seek to impose the structure of western theatre on to everyday social dramas. On the contrary, Turner sees that social drama 'is a well nigh universal processural form' (71) and with this, turns the privileging of aesthetic drama on its head. From this perspective, social drama is proposed as a cultural universal which has a particular relation to the 'restored behaviour' (a term Turner borrowed from Schechner 1985: 35) characteristic of western aesthetic drama. Turner considers that there is a 'dialectical relationship between social performances and genres of cultural performances in all cultures' (1982: 72). Like Goffman, whose approach is discussed below, Turner initially incorporated the metaphor of the theatre to examine a range of social processes in life as opposed to the theatre. In this sense, it could be argued that his approach is better situated within the domain of 'social performances'. However, his collaborations with Richard Schechner, the theatre theorist whose work has been of seminal importance to the development of performance theory, led him to a point where he argued that a collaborative approach between anthropology and theatre studies, centring on the concepts of performance and 'drama' 'could become a major teaching tool' (91). In turn, this led him into the hybrid arena of 'performance ethnography' (1982), which is not to be confused with the notion of ethnodrama discussed earlier in the chapter. Moreover, as suggested above, his ideas have found their way into theatre studies as well as anthropology, but he also has his detractors (see Shepherd and Wallis 2004). Although anthropologist Kirsten Hastrup (1995) recognizes the import of Turner's contribution to the field and indeed, draws on his ideas, she nevertheless parts company with him in certain respects. For example, while she recognizes that there are certain qualitative distinctions between theatrical and everyday (social) performances, she rejects the division that Turner came to make between the two modes. In this regard, she also parts company with Eugenio Barba (1991), the founder of 'theatre anthropology' who was also a considerable influence on her work (Hastrup 1992). Turner, it should be pointed out, did include events such as ritual and religious rites, which are not normally seen as 'theatrical' into his performance and drama frame of analysis. Theatrical performances, Hastrup concedes, 'are designed to realize presence by way of illusion', whereas social performances are concerned with 'illuding reality by way of presence' (1995: 78). She also recognizes that 'the human body is enculturated largely implicitly, whereas actors are explicitly enculturated into the conventions of theatre' (ibid.). However, she proposes that taking 'the performing body as the point of departure', allows for the possibility of conflating the two. Adopting the perspective of 'acting body' would facilitate an acknowledgement that 'there is a profound continuity between acting on stage and living in the world' (ibid.). As such, she maintains that 'ordinary and extraordinary bodily experiences and expressions should not be distinguished, as 'both are within the range of the normal' (78, 79).

Turner's framework of social drama, which he first outlined in the late 1950s, is indebted to the ethnologist and folklorist Arnold Van Gennep's (1960 [1909]) model of the 'rites of passage'. Van Gennep categorized the organization of rituals that marked the passage of individuals or whole groups from one status to another (e.g. the rites associated with the passage from childhood to adulthood) in small-scale societies into three processural stages: '*separation, transition and incorporation*' (cited in V. Turner 1982: 24). The first separation phase involves the 'detachment of ritual subjects' or groups in space and time from their everyday 'previous social statuses' (ibid.), so that they are literally and metaphorically spatially set apart and 'out of time'. In the transitional phase, which Van Gennep termed the 'margin' or the 'limen', which means 'threshold' in Latin, 'the ritual subjects pass through a period and area of ambiguity, which has few . . . of the attributes of either the preceding or subsequent profane social statuses or cultural states' (ibid.); this is the betwixt and between state. In the 'liminal' phase, the world is turned upside down, the social order inverted. The third stage 'involves symbolic phenomena and actions which represent the return of the subjects to their new, relatively stable, well-defined position in the total society' (ibid.).

Turner later made a distinction between the liminal and the 'liminoid'. The former is associated with ritual behaviour in traditional or tribal societies, which are founded on the basis of what Durkheim termed 'mechanical solidarity', while the latter is to be found in modern industrial societies, which Durkheim (1984 [1893]) called 'organic solidarity'. While liminal performances may invert the social order, Turner argues that they do not essentially subvert it, whereas liminoid performances are more likely to both invert and subvert, either by accident or intention. As such, the liminoid provides a possible space for social and cultural resistance.

Carlson (2004: 16) notes that Turner's adaptation of Van Gennep's model has had significant ramifications for later formulations of performance theory. While thinkers like Hymes and Bauman, among others, treat performance as being distinct or set apart from the flow of everyday life, Turner is more concerned with 'its "inbetweenness", its function as transition between two states of more settled or more conventional cultural activity' (ibid.). It is this 'image of performance as a border, a margin, a site of negotiation, which paved the way for later thinking on what we mean by performance (ibid.).

The margin, as other thinkers have argued (Bakhtin 1984 [1968]; Douglas 1970; Kristeva 1984), can be a dangerous, creative, playful, potentially subversive or liberating site to inhabit. The subversive character of the betwixt and between is perhaps most associated with Mikhail Bakhtin's (1984 [1968]) discussion of carnival and the carnivalesque. Bakhtin's construct of the carnivalesque resonates closely with Turner's notion of liminality (Carlson 2004: 23) and it has had an impact on contemporary performance theory and body studies, particularly in regard to gendered bodies, as Mary Russo (1986) shows.

Bakhtin (1984 [1968]: 122–3) considers that the 'concretely sensuous, half-real and half-play-acted form' of carnival opens up possibilities for generating new forms of social relations between individuals which are not predicated on the inequalities inherent in the normal structure of society. During carnival, the social distances between people and the social hierarchies experienced in everyday, 'noncarnival' life are suspended. Bakhtin's idea of carnival, according to sociologist Alan Swingewood (1998: 127), is underpinned by a 'utopianism of absolute freedom'. In carnival, the customary rules and prohibitions of the social order have been inverted, 'as carnivalised forms subvert and transgress the official social order'. The powers that are set into motion through the carnivalesque are momentary, unstable and incomplete, providing a free zone for testing the possibilities of new social and cultural orders. As such, Bakhtin's notion of the powers of the carnivalesque closely corresponds with Victor Turner's view of the liminal or the liminoid (Carlson 2004: 23). Like Turner, Bakhtin maintains that there is a distinction between the carnivalized forms of former eras, the highpoint of which was in the renaissance period of early modern Europe, and 'its more mediated, truncated and scattered modern descendants' (ibid.). Bakhtin also notes that 'something of the carnivalistic licence' available in earlier forms is retained in modern theatre and spectacle.

BODY BOUNDARIES AND MARGINS: POWERS AND DANGERS

It was Bakhtin's theory of carnival as a space of possibilities for transforming everyday social relations and social structures through play that resonated with the interests of performance artists and theorists in the late 1980s who were also exploring these concerns through their playful, carnivalesque interventions in the 'society of the spectacle' (Debord 2005 [1967]). Here, once more, the body takes centre stage, particularly in regard to Bakhtin's emphasis on the 'grotesque body'. A number of feminists in particular have drawn on Bakhtin's idea of the grotesque body to explore and disrupt notions of the female grotesque which underlies much cultural imagery and symbolism and which is suffused with the 'connotations of fear and loathing' connected with the biological processes of women's reproductive capacities and of ageing (Russo 1986: 219). Thus, Bakhtin's ideas have extended beyond aesthetic and cultural performances to the arena of social and cultural analysis. As Mary Russo (1986: 214) points out;

> the work across the discourse of carnival or, more properly, the carnivalesque . . . has translocated issues of bodily exposure and containment, disguise and gender masquerade, abjection and marginality, parody and excess, to the field of the social constituted as a system of the symbolic.

Mary Douglas (1975: 56) argues that it is not only marginal social states that must be viewed as dangerous, rather all margins and boundaries, particularly those of the body, are treated as potentially 'dangerous and polluting'. The boundary of the body in Douglas' schema, as Butler (1990: 131) notes, constitutes 'the limit of the social per se'. Douglas maintains that matter and fluids emanating from the body's orifices or edges (sick, mucus, faeces etc.) are potentially dangerous to the social order of a given society and are often surrounded by rituals to curtail the polluting potential of the outflow of fluids from the inside of the (individual) body to the outside (social) world.

'The outflow of the body' as Margrit Shildrick (1997: 17) points out, 'breaches the boundaries of the proper'. The dominant cultural dualistic differences in western thought, 'mind/body, self/other, inner/outer – are threatened by a loss of definition, or by dissolution' (ibid.). In 'the male cultural imaginary', she goes on to suggest, following the post-structuralist feminist Luce Irigaray, the 'metaphor' of 'fluid boundaries' 'characteristically provokes unease, even horror' (ibid.). The 'sex panic' (Kroker and Kroker 1987), which followed the classification of AIDS as a killer disease in the late 1980s, offers a clear example of this. AIDS was labelled by the hysterical mass media as the 'gay disease' and the individual who contracted the illness was constructed as a 'polluting person' (Watney cited in Butler 1990: 132). In this homophobic discourse, as Butler notes, 'there is a continuity between the polluted status of the homosexual by virtue of the boundary-trespass that *is* homosexuality and the disease as a specific modality of homosexual pollution'; which is that it is transmitted through 'the exchange of bodily fluids' (ibid.). The homophobic figuring of this double transgression suggests 'the dangers that permeable body boundaries present to the social order' (ibid.).

In Kristeva's psychoanalytic feminist framework (1984), the margin represents the limit of patriarchal symbolic power, the 'law of the father' in Lacanian psychoanalytic theory, which is constituted through the rules of language, beyond which chaos and the danger of madness lie. Because the margin shares the 'properties of all frontiers', Kristeva considers that those who occupy that border 'will neither be inside or outside, neither known or unknown' (Moi 1985: 167). Whereas patriarchal symbolic power is constituted through language, the feminine other, for Kristeva, is articulated in the prelinguistic 'semiotic' phase (prior to the child's entry into language), which she calls the 'chora' (1984: 24). It is important to note that the construct of the feminine in Kristeva's schema does not refer to biological femaleness. Rather, following Freud, she considers that human beings are polymorphously perverse at root and that both women and men can potentially embody the feminine. The chora is made up of bodily 'drives' and 'pulsions'; it is a mobile kinetic force that comes from the body and is rooted in the prelinguistic phase. With the entry into language, the semiotic and thus the feminine (pre-Oedipal maternal body or the 'imaginary' in Lacanian theory), is repressed and pushed to the margins of the symbolic order, where, because it cannot be wholly contained by the rational structure of language, it has the potential to disrupt, challenge and threaten the stability of phallocentrism.

> Kristeva proposes that it is in the writing of the modernist literary avant-garde such as that of Joyce or Mallarmé that the repressed feminine finds a voice, where the bodily 'pulsations' and the unconscious break through the rational barriers inscribed in the phallocentric order.
>
> (Thomas 2003: 169)

Kristeva, then, emphasizes the potentially subversive and dangerous site of the margin; what lies beyond the margin is chaos and abjection. As Butler points out in her discussion of Kristeva's *Powers of Horror* (1982), 'the abject designates that which has been expelled from the body, discharged as excrement, literally rendered "Other" ' (1990: 133).

The above truncated, selective discussion on the uses of performance has focused on aesthetic or cultural performances and the ways in which performance has been used and developed within this domain. However, as with the brief outline of Turner's approach and the discussion that developed from the idea of the 'betwixt and between', it should by now be clear these performative practices find their way into the world of everyday life, for want of a better term, and that the influence of certain theoretical and methodological frames straddle performance studies and social and cultural studies. This perhaps reflects the border crossings and the dissolution of disciplinary boundaries which have taken place since the 1980s in more established social science disciplines like sociology and anthropology. Newer, related subject areas such as gender studies, performance studies and cultural studies, it should be noted, were constituted through interdisciplinarity from the very beginning. In the section that follows, the discussion turns more centrally towards the second usage of performance in the social sciences. Here, the focus veers more towards the notion of performativity, which, as we shall see, also overlaps with a number of issues touched upon in the discussion of aesthetic and cultural performances.

SOCIAL PERFORMANCES

The early work of Erving Goffman, particularly *The Presentation of Self in Everyday Life* (1971 [1959]), which championed a dramaturgical model, is usually linked with the second usage of performance, which refers to social performances, as indicated earlier in the chapter. Here, the theatrical metaphor is brought to bear on activities which are situated in the 'real world' (for want of a better term) of everyday life, as opposed to a theatrical performance which takes place in a marked out, bounded space, as outlined above. Although, as I suggested earlier, this notion of a fixed, bounded frame is also subject to question.

In *The Presentation of Self in Every Day Life*, Goffman outlines and demonstrates what he sees as 'six dramaturgical "principles"' at work in real life everyday performances involving social actors: 'the performance, the team, the region, discrepant role, communication out of character, and impression management' (Manning 1992: 40). Goffman begins from the principle that 'when an individual appears before the presence of others he will have many motives for trying to control the impression they receive of the situation' (1971 [1959]: 26). His 'report', as he calls the book, is concerned with a range of 'the most common techniques' that individuals employ to maintain the impressions they give out and 'some of the common contingencies' with which these techniques are associated (ibid.). Goffman focuses on the 'dramaturgical problems' that individuals encounter when presenting their 'activities' to others in 'face-to-face' interactions (ibid.).

Goffman's rationale for drawing on a theatrical model is based on the fact that the 'general' and sometimes 'trivial' issues which 'stage craft and stage management' routinely deal with 'are everywhere in social life and as such, are suitable for formal sociological analysis' (ibid.). The theatrical metaphor is also a heuristic device for social analysts to defamiliarize themselves from what is known or taken for granted as they immerse themselves in the minutiae of everyday life.

Although people generally do not consciously think about their everyday performed actions on a moment-by-moment basis as they, say, walk down the street, Goffman reveals such practices as 'a set of implicit instructions' (Manning 1992: 5) that governs behaviour. Goffman explored these bodily performances in public places in greater depth in two of his subsequent works (1963; 1972) and he later examined representations of gendered bodily expressions and gestures in advertising (1979). Goffman's analysis reveals that slight departures from these business-as-usual rules can result in an undermining of our confidence in the social world as it appears to us on a routine basis. Myriad concrete examples that Goffman uses are designed to show those features which underpin our self-assurance in the everyday social world in which we live. The taken for granted actions in everyday life are unpacked to become 'quite alien and startling' (Manning 1992: 5). In a sense, this device facilitates the possibility of looking at everyday actions and situations with a different pair of eyes through a process of defamiliarization.

In Goffman's world of (theatrical) social performances involving 'impression management' and 'team performances', the individual is portrayed as 'a cynical role-player who hides behind an array of masks' (8). In his dramaturgical model (a model which altered and developed in relation to his subsequent work), according to Tom Burns (1992: 106–7), we are confronted by 'a series of selves, one inside the other', rather like a 'Russian doll', underneath which there is also an inner 'I' which '*manages*' the social self. The legacy of the symbolic interactionist tradition emanating from George Herbert Mead (1934) is evident here. In Mead's theory of the self, the individual subject is split between the 'I' and the 'me'. The 'me' represents the social self constituted through the individual's interaction with significant others, while the private 'I' comprises the inner, true self of the individual.

> There is an inner self lurking inside the self that is present, or presented, to the outside world of others. The divisions match those between playwright, producer, actor, and part. There is a social self ('producer') which measures the appropriateness of the individual's role to the social position in which it is fixed ('part'), and also adjusts the distance between them – i.e. the degree to which it seems rewarding to measure up to performance of the role, at its most typical ('actor').
>
> (Mead 1934: 107)

The idea that 'all the world is a stage', of course, is an old one; it can be found in the writing of classical Greece (Plato), the renaissance (Petronius) as well as in Shakespeare. In the twentieth century, the notion that the social world is a stage, with actors performing a play within it, emerged in the late 1940s and 1950s in relation to 'social role' theory (109–10). Social scientific interest in examining 'social roles' and social 'actors' was given impetus by anthropologists such as Marcel Mauss and later Victor Turner. However as noted earlier, the latter, unlike Goffman, maintained a distinction between performing in everyday life and performing on the theatrical stage: between social and aesthetic dramas, although he viewed them as interconnected. Interest in social roles/ actors was also reflected the development of humanist concerns in Britain, France and the US in the period between the two World Wars, which then burst on to the social science scene in the post-1945 era. Goffman, for example, was influenced by Kenneth Burke's *Grammar of Motives* (1945). Burke invoked an approach which he termed 'dramatism' to study the vocabulary of motives from a viewpoint taken from the model of drama, 'which treats language and thought as modes of action' (Burke cited in Burns 1992: 108).

Goffman defines a (social) performance as 'all the activity of a given participant on a given occasion which serves to influence in any way any of the other participants' (1971: 26). As such, according to Carlson (2004), it could accommodate much of the artistic activity that has gone under the title of performance in recent years. Goffman does not necessarily lay stress on the social actor's agency, as is normally evident in theories of role play and presentation. Rather, for Goffman, as Carlson notes, it is the presence of an audience that 'makes a performance a performance and not simply behaviour' (35). The audience, in this sense, has an effect on the performance or behaviour of the performer; the audience 'acts' back, and is not simply passive. Moreover, in a Goffmanesque performance, the individual may in fact be performing without being aware of it.

Goffman's approach draws attention to the actions and the functions of the performer and the audience. Like most sociologists, however, he stresses the latter over the former; 'the community in which performance occurs' (ibid.). It should also be noted that his consideration of the individual 'performing' without consciously so doing resonates strongly with later social and cultural approaches to the body, such as that of Butler (1990, 1993) and Bourdieu, who was influenced by Goffman's ideas on the 'infinitely small' (1983), and 'total institutions' (1977).

Bourdieu, like Goffman, is centrally concerned with social practices. Whilst Bourdieu is interested with what individuals *do* in their everyday lives, he nevertheless insists that practice is not to be comprehended on the basis of individual (subjectivist) actions or resolve alone, nor is it wholly determined by external (objectivist) social structures. As Richard Jenkins points out, Bourdieu's fine tuning and application of the concept of habitus 'is a bridge-building exercise across the explanatory gap between these two extremes, another important device for transcending the sterility of the opposition between objectivism and subjectivism' (1992: 74). This approach will be discussed further in Chapter 4 in relation to Wacquant's (2004) ethnographic study of boxing.

'Habitus' is a Latin word that literally 'refers to a habitual or typical condition, particularly of the body' (Jenkins 1992: 74). In Bourdieu's work (1984), the concepts of habitus and its related notion of 'hexis' (gestures, postures, gait etc.), both of which are embodied, are perceived to be relatively permanent, stable, bodily 'dispositions', which bear the marks of social class, status and gender and we may suggest, other markers of social difference. While the construct of habitus is close to the notion of 'habit', it is distinguished from it in a key aspect. 'The habitus . . . is that which one has acquired but which has become durably incorporated in the body in the form of permanent dispositions' (Bourdieu 1993: 86). As such, it is incorporated into 'individual history' which means that it is not an 'essentialist' form of attention. Bourdieu considers that the habitus is a 'capital' (a property) but 'because it is embodied, [it] appears as innate' (87). It differs from the idea of habit, which is 'regarded as repetitive, mechanical, automatic, reproductive rather than productive' (ibid.). Rather, for Bourdieu, the habitus 'is something powerfully generative'; it is both reproductive and productive. Somewhat echoing Marx's notion of history in reverse (made by men [sic] but not of their own choosing), Bourdieu sees that the habitus 'is a kind of transforming machine that leads us to "reproduce" the social conditions of our own production, but in a relatively unpredictable way' (ibid.).

Goffman's attentiveness to the 'infinitely small' aspects of everyday life is recognizable in Bourdieu's attention to bodily hexis, as he (1983) readily acknowledges (see also Jenkins 1992). Bourdieu defines bodily hexis as 'a political mythology realized, *em-bodied*, turned into a permanent disposition, a durable manner of standing, speaking and thereby *feeling* and *thinking*', which are 'beyond the grasp of consciousness' (1977: 93–4). He draws on his ethnographic study of the Kabylia to show how the 'mythico-ritual logic' of the oppositional value system between men and women is revealed in 'the gestures and movements of the body, in the form of opposition between the straight and the bent, or between assurance and restraint' (94). The male ideals of assurance and honour are revealed in the men's upright stance, decisive gait and firm, outward, forward orientation towards other men and the world. By contrast, the female values of restraint and modesty are shown in the women's slightly bent, stooped deportment, which is accompanied by a fixed downward gaze towards the ground where she is walking. Bourdieu's bodily hexis, as Jenkins (1992: 75) points out, 'combines the idiosyncratic (the personal) with the systematic (the social)'. As such, bodily hexis becomes the 'mediating link' between the subjective world of the individual and the social world into which they are born and inculcated into along with others (ibid.).

Butler does not draw directly on Goffman's ideas in either of her two major books which led to a mini industry in challenging the conventions of 'gender performativity', although it is clear that she does in certain respects follow Bourdieu's line, while departing strongly from it (see Butler 1999 on this). However, her concern with the ways in which gender normativity is routinely played out in men and women's performative bodily acts and gestures, without them consciously attempting to so do, resonates strongly with aspects of Goffman's ideas on social performances, albeit with a decidedly deconstructive edge. Indeed, for Butler, in her earlier work at least (1990), when gender identity and thus the notion of true biological sexual differences are revealed as fictions through such bodily transgressions as 'drag', then performativity is revealed through performance. As such, performance practice such as drag provides the possibility of challenging the unquestioned hetero-normative discourses that are inscribed on the body and yet treated for all practical purposes as expressions of biological binary sexual difference. The 'performative gendered body' (through acts, gestures and desires), on the other hand, generates 'the illusion of an interior and organizing gender core, an illusion discursively maintained for the purposes of the regulation of sexuality within the obligatory frame of reproductive heterosexuality' (136), as will be discussed more fully in the following chapter.

With a Goffmanesque approach, as Schieffelin (1998) points out, the centre of attention is not on a 'type of performance event', as is evidenced in the first aesthetic and cultural usage discussed earlier in the chapter. Rather, the focus is on performativity: 'the expressive processes of strategic impression management and structured improvisation through which human beings normally articulate their purposes, situations and relationships in everyday life' (ibid.). The construct of 'performativity', which has gained considerable currency in cultural writing since the 1970s, owes much to attempts by linguistic philosophers to show that 'speech' is not simply saying or reporting on something but rather, it could be said to be *doing* something.

PERFORMATIVITY

The concern to view certain kinds of speech as not just saying something but doing something, as actions, was advanced by the Oxford philosopher J. L. Austin in *How to Do Things with Words* (1962). Austin was concerned with everyday speech acts as social practices, just as Goffman was concerned to show the unquestioned social practices that underpin in everyday performances. In so doing, Austin drew attention to a particular kind of utterance which he labelled a 'performative'. For example, in voicing the words, 'I name this ship Queen Elizabeth', or saying, 'I do', in a wedding ceremony, we are not simply making a statement, according to Austin (1962: 222), rather we are *doing* something; we are performing an action. From this viewpoint, words act and they enact what they name. The measurement of the success of such performative statements is to be found in whether they succeed or not (their 'felicity' or 'infelicity' in Austin's terms), rather than on their truth or falsity.

Austin's 'speech act' approach was in sharp contrast to the dominant model of linguistics of the 1960s and 1970s, Noam Chomsky's (1965) 'transformational-generative grammar'. Chomsky's linguistic model separated competence and performance, somewhat along the lines of de Saussure's (1974 [1915]) distinction between 'langue' (language) and 'parole' (speech), although with significant differences. Everyday speech acts (performance) were deemed to lie outside of the frame of linguistic theory, which, he maintained, should be concerned to understand the underlying language structure (competence), which was made manifest through speech (performance). In Chomsky's model, the underlying rules of language are the most important and the performance part of speech merely provides the instance for unveiling the grammatical rules.

'The general term that Austin began to apply to his concerns was "illocution" ' (Carlson 2004: 62). He subsequently sub-divided speech acts into three types: 'locutionary', 'illocutionary' and 'perlocutionary'. An illocutionary speech-act refers to 'utterances with a certain conventional "force"; they call into being, order, and promise, but also inform, affirm, assert, remark, and so on' (ibid.). Austin's notion of illocutionary is useful in terms of thinking through ways in which we might treat certain bodily expressions or actions as performative, as the quotation from Schieffelin at the beginning of the section on 'uses of performance' in this chapter would seem to indicate. This would require transposing the notion of a speech act into a body act. In this instance, illocutionary would suggest both a summoning up of something of itself and a reflecting and commenting back on itself, in and through a body-act. Les Back's (2004) photographic essay on bodily inscriptions provides a thought-provoking example of this.

Back uses a case study approach to examine the ways in which bodily inscriptions in the form of tattoos can tell 'stories', which speak to and of, matters which are not usually spoken or heard in particular communities. He does not focus on the spectacular, consumerist, neo-primitive tattooing or body modification revival of recent years, which gained much academic attention in the first few years of the millennium in social and cultural studies (Featherstone 2000; DeMello 2000; Pitts 2003; Krakow 2005), although he draws on more socio-historical studies (for example, Caplan 2000). Rather, his individual case studies are drawn from more locally based white working-class communities in south-east London. Back (2004: 40) explores the intricate ways that memory, place, class and love are called forth and are 'brought into being without pains-taking announcement' by focusing on this historically denigrated form of inscription in western culture.

The act of tattooing, as Back points out, is a corporeal experience. The tattoo itself, which is inscribed on the individual's body or perhaps more accurately speaking, pierced into and through the flesh, constitutes an illocutionary act. Back explores the ways in which the tattooed image can be seen as performing an act of love for someone or something dear to the heart of the tattooed person. Crucially, for Back, it is through the tattooed image engraved on the body that the traditionally voiceless 'stories' of these working-class people 'speak' and may be heard, rather than through verbal utterances. This performed love is 'written on and in the body by tattooing: the expression of love of a football fan and a locality; a father for his beloved children, a grand-daughter for her parents and her familial past and an older son for his recently demised father' (Thomas and Ahmed 2004: 9). For example, Back draws on his brother's actions in relation to his father's death to demonstrate an illocutionary love that performs its name through embodied practices; 'a love that was rarely, if ever, brought into speech' between father and son. When his father died, Back's older brother, Ken, began for a time to 'inhabit his father's absence in a literal way' by wearing his clothes, 'even his glasses' (Back 2004: 46). In so doing, he also appeared to take on his father's gestures and movement; calling forth the absent-presence of his father through his own body. Ken also had a tattoo inscribed on the top of his left arm; 'a graphic imago, consisting of swallow in flight, holding a scroll in its beak, and on the fleshy parchment was inscribed three letters – DAD' (47). This tattoo, for Back 'names the object of illocutionary love', which is permanently inscribed on Ken's body (ibid.), a love that was never really voiced in words between father and son, which, as he suggests, is often the case in working-class households.

As Back notes, however, while tattoos may offer a glimpse of permanence, they cannot remain intact forever. Despite the notion within consumer culture that the body can be made and remade at will, the tattoos and thus, the illocutionary love these 'working-class' inscriptions called into being, will disappear with the death of the owner and therein lies the paradox.

> [T]his has a double consequence for working-class expression because it is often the only medium through which their stories are told. There is no place for them in what Jacques Derrida calls a 'hospitable memory' . . . As the cadavers disappear, the traces of their embodied history, of life and love, are lost – they become The Nameless. They pass through hospital wards to the crematoria, to be remembered in the inscriptions made on the young flesh that will in turn grow old.
>
> (Back 2004: 51)

Austin's focus on performative utterances gained much currency in linguistic anthropology in the 1970s (Bauman 1992) and it 'has resonated through a range of theoretical writings over the past three decades' as Parker and Sedgwick (1995: 1) note. Although Austin sought to separate the stage 'actor's citational practices from ordinary speech-act performatives', Derrida would later argue that there was a 'pervasive theatricality common to the stage and world alike' (2). Thus, Derrida came to challenge the notion of the separation of the stage and the social world. As indicated earlier, the re-reading of Austin by both Derrida and Butler has been important in the renewed postmodern interest in performativity in a range of discourses such as literary studies, performance studies and feminist, gay and queer studies, which have often provided a productive crossing of boundaries between performativity and the 'loose cluster of theatrical practices, relations and traditions known as performance' (1).

CONCLUSION: BODY, PERFORMANCE, SOCIAL ACTION

The two main approaches to performance or performativity outlined in the discussion of aesthetic and social performances generally 'draw inspiration and conceptual terminology from the western theatrical performance tradition'. Schieffelin's (1998) essay on performance in social science, cited earlier in the chapter, is concerned to explore the advantages for the area of study in drawing on theatrical concepts and also the limitations of so doing. He considers that 'there is something fundamentally performative about human being-in-the-world' (195). He seems to be proposing a kind of universal human expressivity in a similar vein to the way in which Mary Douglas (1973), some years before, argued that there appears to be a human propensity for bodily expressivity across cultures. Schieffelin proposes that 'without living human bodily expressivity, conversation and social presence, there would be no culture and no society' (1998: 195). He further suggests it is 'because human sociality continues in moment-by-moment existence only as human purposes and practices are performatively articulated in the world that performance is (or should be) of fundamental interest to anthropology' (ibid.) and we might add, to sociology. However, he considers the western theatrical tradition on which much of the work on performance in the social sciences is based, can be problematic if unwittingly applied as a general theory of action to the performative practices in other cultures. This is because the western dramaturgical model is founded on the notion of 'illusion' and as such, it offers a model of action that is based on a sleight of hand (think of Goffman's idea of manipulation here), 'the fake, the unreal, the simulacrum' (Hughes-Freeland 1998: 13). It is thus tied into a series of binary opposites; performer/audience, 'deceit/authenticity, activity/passivity' (ibid.), which are endemic in the western humanist tradition. Therein sits the danger of ethnocentrism, of analysing 'other' worlds through the prism of the privileged position of one's native cultural configurations. What is at stake in the ethnography of performance, for Schieffelin, is the 'social construction of reality'; that is, 'the imaginative creation of a human world' (1998: 205). This requires an exploration of theories of knowledge which are not constructed before the investigation in terms of some preordained relationship between, for example, performer and spectator in formal and social performances, based on the western model of dramaturgy. This does not mean that performance practices should be kept separate from ordinary every day social practices. Indeed, he suggests that it is within the relationships between formal performances and social performances that 'fundamental epistemological and ontological relations of *any society* are likely to be implicated and worked out' (204, my emphasis). Just as 'western models about the relation between performer and spectator convey fundamental (western assumptions) about the nature of actions in ordinary situations' (ibid.), we could hypothesize that the same would be true for other cultures or other historical eras, although they may be based on different assumptions and have different implications. Thus, he proposes that there is a rich creative edge in the relationships between formal and social performances, through which we might begin to come to grips with the 'social construction of knowledge'.

3

PERFORMING GENDER: THE 'BODY' IN QUESTION

INTRODUCTION

The previous chapter considered a range of theories and practices captured underneath the headings aesthetic, cultural and social performances, in which ideas concerning the body are closely implicated and entwined. There, I argued that the distinctions between the aesthetic and the everyday are not clear-cut and that in significant ways, these theories and practices flow into and inform each other. In this chapter, I extend that discussion to focus on the ways in which gender is constructed and produced in specific kinds of representations of women's bodies. The 'body' in question, here, is largely centred on shifting ideas about women's relations to their bodies in feminist discourse, which, as indicated in Chapter 1, contributed to the increased interest in the study of the body in social and cultural studies. In this chapter, the changing theoretical frameworks of feminist discourse in regard to the distinctions between nature and culture, sex and gender, and femininity and masculinity, are explored in more detail. These themes are later raised in regard to plastic surgery (reconstructive and cosmetic), which has become a topic of feminist debate and cultural criticism in recent years. Whilst plastic surgery is generally considered to be a practice that was piloted and subsequently normalized in the US, there is sufficient evidence to show that the demand for cosmetic surgery among women in particular, although not exclusively, has grown exponentially in recent years in the UK and Europe.

The American Society for Aesthetic Plastic Surgery (ASAPS) has provided detailed annual statistical information on surgical and non-surgical procedures performed by its members since 1997. The ASAPS reported that there were almost 11.5 million surgical and non-surgical procedures performed in 2005–2006, with women accounting for 92 per cent of these (http://www.surgery.org/media/news-releases/115-million-cosmetic-procedures-in-2006). While surgical procedures represented 17 per cent of the total, non-surgical procedures accounted for 83 per cent. Overall, surgical procedures were down by 9 per cent, while non-surgical procedures increased by 4 per cent on the previous year. The top surgical procedure was liposuction for both men and women and the most popular non-surgical (or minimally invasive) procedure was Botox injection. The 2010 ASAPS survey reported that over 9.3 million surgical and non-surgical procedures were performed in 2009–2010 (http://www.surgery.org/sites/default/files/Stats2010_1.pdf). The ASAPS press release states that despite a downturn the number of procedures performed in the few years before, probably due to 'uncertainty in the economy and the job market', there had been a 9 per cent increase in plastic surgery procedures for that year (http://www.surgery.org/media/news-releases/demand-for-plastic-surgery-rebounds-by-almost-9percent). The percentage difference between surgical and non-surgical procedures remained the same as in the 2005 survey, as did the percentages of women and men; 92 per cent women, 8 per cent men. Liposuction remained the most popular procedure for men throughout this period. However, since 2008, breast augmentation has replaced liposuction as the most favoured procedure for women.

The British Association of Aesthetic Plastic Surgeons' (BAAPS) annual audit for 2005 reported that 22,041 surgical procedures were performed, which represented 34 per cent increase on the previous year (http://www.baaps.org.uk/about-us/audit/49-over-22000-surgical-procedures-in-the-uk-in-2005). While the overwhelming majority of procedures were carried out on women, 89 per cent, there was a 3 per cent rise of procedures on men compared with the previous year. The latest figures released by BAAPS show that 38,274 surgical procedures were performed in 2010, which represents a slight rise of 5 per cent overall from the previous year (http://www.baaps.org.uk/about-us/audit/854-moobs-and-boobs-double-ddigit-rise). Ninety per cent of surgical operations were carried out on women, with breast augmentation, up by over 10 per cent on the previous year, remaining the most favoured procedure.

However, the audit shows that there was a 7 per cent rise in men undergoing cosmetic surgery compared with the previous year.

The BAAPS statistics since 2004–2005, although not as comprehensive as those of the ASAPS by any means, show that cosmetic surgical procedures have more than doubled in the UK during this period. Moreover, there is a recognizable growth in men's interest in plastic surgery in the UK, with rhinoplasty or 'nose jobs' remaining the most popular. There was an increase of 28 per cent in gynaecomastia or 'man boob' procedures on the previous year, making it the second most common operation for men for the first time. Although these UK numbers pale in significance in comparison with the US, there are more surgical procedures carried out in this country than the rest of Europe. The figures for Europe in Frost & Sullivan's 2008 market trends survey also indicate a significant growth in this market (http://www.frost.com/prod/servlet/market-insight-top.pag?docid=153913646).

This surge towards body re-modification seems to point to a cultural shift in our everyday perception of the body, which is increasingly viewed as a pliable substance that can be altered and remodelled via the surgeon's knife; frozen in time with non-surgical procedures such as Botox, or other non-invasive or minimally invasive, anti-ageing procedures and products. Ideas around anti-ageing will re-surface in Chapter 5. For a broadly based discussion of surgery and the body, see Julie Doyle and Katrina Roen's Introduction to the *Body & Society* special issue, Surgery and Embodiment (2008).

The aim of this chapter is to weave the discussion of changing perceptions of bodies around two 'performance' case studies in particular. The first is the photographer Jo Spence, focusing on her autobiographical photographic and written work which charts her breast cancer pre and post-operation and the development of phototherapy in the 1980s. The second is the performance artist, Orlan, paying particular attention to her cosmetic surgical performances in the 1990s, as part of what she terms her approach to 'carnal art'. I will suggest that these two approaches speak to two different moments in feminist discourse, the first encapsulating key elements of second wave feminist approaches and the second, postmodern feminist approaches to the body. This does not mean, however, that the work of these two women cannot be read in different ways. Indeed, I suggest that their work also gestures to different theoretical perspectives and can be viewed on several levels. One of the problems with analysing the impact of Orlan's work in the 1990s, for example, is that it seems to be postmodernist and feminist, with elements of modernist (art) tendencies embedded in it, whilst it also appears to be moving towards a posthumanist perspective. It has also been subject to critical review from a range of academic discourses, for example, feminist and postfeminist perspectives on cosmetic surgery (Davis 1997; Blum 2003), a psychoanalytic viewpoint (Adams 1996), art history (Rose 1993; Wilson 1996; Ince 2000) and the history of aesthetic surgery (Gilman 1999), as well as the international mass media. Jo Spence, on the other hand appears to be firmly rooted in a form of socialist feminism but her later work suggests a movement towards a more postmodernist type of approach in some ways. Before this, however, it is necessary to make clear some of the key differences between these two strands of feminism, while bearing in mind that there are in fact different feminist viewpoints within and across these positions. As an introduction to this, I begin by considering the quotation below from a chapter on 'the male body' by a leading American writer, which, by implication, also has something to say about the female body.

FROM GENDER TO SEXUALITY

> Inhabiting a male body is much like having a bank account; as long as it's healthy, you don't think much about it. Compared to the female body, it is a low-maintenance proposition: a shower now and then, trim the fingernails every 10 days, a hair cut once a month. Oh yes, shaving — scraping or buzzing away at your face every morning.
>
> (Updike 1994: 8)

While the above statement by John Updike on the relative lack of awareness that men generally have in regard to the care and attention of their bodies in everyday life may be a little out of date for young men in particular in contemporary western consumer culture, he nevertheless points to the fact that this is not the case for women. Rather, the female body, by implication is a high maintenance proposition; a subject on which second wave and postmodern feminism has had much to say. Feminist studies on weight and eating disorders (Orbach 1978; Chernin 1986; MacSween 1993; Bordo 1993), the beauty myth (de Beauvoir 1972 [1949]; Friedan 1992 [1962]; Coward 1984; Wolf 1991), body maintenance (Schultze 1990; Johnston 1998), cosmetic surgery (Davis 1995, 2003; Fraser 2003; Blum 2003), representations of women in the media and culture (Parker and Pollock 1987; Kappeler 1986; hooks 1992; Goldstein 1991; Nead 1992), as well as issues around rape, abortion, reproductive technologies and women's health (Oakley 1979; Martin 1987), testify to the perceived close association of women to their bodies, or perhaps more precisely with perceptions of 'the body' in the western cultural tradition.

The 'female body', as Margaret Atwood (1991: 1) notes, has become a 'hot topic', however, she also questions the idea of a singular female body by pointing out that in everyday life 'there is a wider range' of bodies on view, including her own. Questions of mind versus body, nature versus culture, determinacy versus agency, male versus female and sameness versus difference, are never very far away from the surface in many of these discussions on women's bodies. In other words, the work engages with and more often than not, seeks to challenge the dualistic character of western cultural thought (see Bordo 1993). However, as postmodern and poststructuralist writers have argued, these challenges have often unwittingly ended up by securing the binaries (Nicholson 1990; Butler 1990; Gatens 1996) and thus, the dominant cultural imperatives they seek to undercut. Elizabeth Grosz (1994: 3), for example, argues that 'Feminism has uncritically adopted, many philosophical assumptions regarding the role of the body in social, political, cultural, psychical, and sexual life and, in this sense at least, can be regarded as complicit in the misogyny that characterizes Western reason'. Before going on to take up some of these issues in more detail, I want to return to Updike's view of the male body's performance.

Updike's (1994) discussion of the distancing of men from their bodies, as indicated above, speaks volumes about presumed distinctions between men and women's approaches to their bodies, which for Updike, appear to be rooted in the 'natural' differences between male and female. Young men are more reckless, more daring in terms of the desire to 'take your body to the edge, and see how it flies' (9). Female space is more 'internal', 'active' and 'interesting' than male space, which has an outer focus – 'the ideal male body is taut with lines of potential force' (ibid.). This description somewhat resonates with Bourdieu's account of the Kabylia males' bodily hexis discussed in the previous chapter. While women appear to have one (maternal) body, men have two, with the penis, particularly during an erection, 'appearing to be only partly theirs' and it even gets its own name, which individuates it further. For Updike, to live in the male body is to 'feel somewhat detached from it'; it is neither friend nor foe. The surface (male)/depth (female) dichotomy is due to the fact that the female body participates more in the processes of nature (presumably through menstruation and giving birth) than the male body. He also notes that in ageing, the male body 'does not betray its tenant as rapidly as a woman's'. By implying that women's bodies age earlier than men's, he shows up the presumed close relation of women to the 'look' in the cultural imaginary.

Much of Updike's take on the distinctions between men and women in relation to their bodies written at the end of the twentieth century would have made second wave gender feminists who were writing in the 1970s and 1980s weep because of the way in which this American man of letters still collapses cultural differences between men and women into the realm of the 'natural'. Although Updike, to some extent, tips his cap in the direction of culture, his discussion overwhelmingly reduces cultural distinctions (for that read learned) between men and women into natural differences (for that read biological) between the male and female body. It is precisely this kind of elision that second wave Anglo-American feminists, particularly socialist feminists, attempted to overcome in the 1970s and early 1980s (Oakley 1974; Barrett 1988), with considerable success in regard to feminist theory, although psychoanalytic approaches were soon creeping in (Brennan 1989). As indicated in Chapter 1, these feminists sought to challenge the 'biology as destiny' viewpoint which had littered western social and political thought for several centuries, and which had served to justify the subordination of women in social life. By maintaining that the observed differences between men and women are a result of socialization, not biological or physiological differences between the sexes, they advocated the privileging of the category of 'gender' over 'sex' as the organizing principle in their analyses of 'the social, familial and discursive construction of subjectivity' (Gatens 1996: 4). Femininity and masculinity, from this perspective, are social constructs, which are not dependent on biological differences between the male and female of the species and therefore, are subject to change over time and place. Under these circumstances, equality between men and women becomes the ultimate goal. Hence, gender feminism is closely associated with the politics of equality, which infers a reduction of difference in favour of sameness through the processes of socialization and/or re-education.

As well as challenging the biological reductionism which, it was perceived, lay at the heart of western social thought from Aristotle through to Durkheim, gender feminists were doing battle with other feminists who proposed that women are closer to nature because of their biology. Certain radical, liberal and existentialist feminists argued that women's biological body processes did indeed contribute to patriarchal oppression and that women had to free themselves from the burdens placed upon them by their bodies in order to gain equality (de Beauvoir 1972 [1949]; Firestone 1970). The social constructionist approach, which was taken up by gender feminists, was thought to stand in sharp opposition to the foundationalist approach adopted by those feminists who argued that biology constituted the basis of women's oppression in society. Simone de Beauvoir (1972 [1949]) was somewhat contradictory in this matter of nature versus culture or foundationalism versus constructionism. On the one hand, she clearly stated that 'one is not born a woman, one becomes one' (301). De Beauvoir maintained that there is no natural essence which constitutes woman per se; rather the real differences between men and women are rooted in socio-historical experiences. She also considered that in order to gain equality with men, women had to transcend their state of being in the world or forever be trapped in their 'immanence'. However, as Iris Marion Young (1998: 260) demonstrates, when de Beauvoir discussed 'women's bodily being and her physical relation to her surroundings', she generally focused on 'the more evident facts of a women's physiology', thus ignoring the 'situatedness of the woman's actual bodily movement and orientation to its surroundings and the world'. In so doing, according to Young, de Beauvoir implied 'that it is woman's anatomy and physiology *as such* that at least in part determine her unfree status' (ibid.). So we can begin to see how de Beauvoir could be categorized as a social constructionist feminist on the one hand and a foundationalist or essentialist feminist on the other. Certain radical second wave feminists, however, viewed the female body in a more positive frame, as a source of wisdom to be celebrated, despite the fact that it had been undervalued and denigrated by patriarchy's denial of and mastery over, nature (Daly 1978; Griffin 1978).

Gender feminism, as indicated above, gained considerable currency as the 1970s progressed. The sex/gender distinction was taken up by several different interest groups including 'Marxists (usually male), some homosexual groups and feminists of equality'. Gatens (1996: 4) argues that despite their different theoretical and political concerns, the inevitable consequences that follow from taking up the sex/gender distinction are 'to encourage a neutralization of sexual difference and sexual politics'.

Feminists who adopted a foundationalist approach based on sexual difference were generally criticized by gender feminists as essentialist in their thinking, universalist in orientation, and therefore, ahistorical. However, as Gatens shows convincingly, one of the key problems with gender feminism is that it took the female and the male body to be passive and inert, a blank sheet ready to be written on through the processes of socialization and in so doing, maintained the mind/body opposition which is the cornerstone of western thought. As a consequence, as Grosz (1994: 4) has argued, social constructionist feminist approaches, by implication, posited 'a biologically, fixed, and ahistorical notion of the body'. Thus, according to Gatens (1996), gender feminism unwittingly operated along the lines of liberal humanist thought, which it sought to undercut. Nevertheless, as I have discussed elsewhere, for quite a long period of time: 'Social constructionism held more sway in Anglo-American feminism and as a consequence, the majority of feminist studies addressed the ideological or cultural trappings of femininity and masculinity and their ramifications for women in contemporary western culture' (Thomas 2003: 36).

Part of the problem lay in the fact that, despite intentions to the contrary, feminist theory displayed a persistent configuration from the late 1960s to the 1980s, which came to be recognized as one of globalizing exclusivity, despite concerns to the contrary. In the 1980s, black feminists and 'women of colour' began to criticize second wave feminist theory for erroneously basing its analysis of the subordination of (all) women in society on white, middle class, North American and western European women. In so doing, feminist theory overlooked or erased the impact of race, class and place on the politics of identity and difference (Carby 1982; Bhavani and Coulson 1986; Minh-ha 1989; hooks 1992, 1994; Brah 1996). Feminists, as Linda Nicholson (1990) has pointed out, began to realize increasingly that they had not only silenced women's voices, but they had also erased the voices of a whole range of different social groups.

With the fragmentation of feminist positions, it was no longer viable to speak of specific kinds of feminisms. The questioning of gender feminism stemmed from a range of sources and involved both pragmatic and theoretical concerns (Gatens 1996). For example: the collapse of Marxism as a viable alternative strategy for the transformation of the social structure; the rise of a different kind of politics of sexual difference through the influence of French psychoanalytic deconstructionist feminist theorists, such as Luce Irigaray (1985), on Anglo-American feminism; the influence of Foucault's (1977; 1984) radicalizing approach to the history of the body and sexuality; the influence of postmodernist theory on feminist thought, and Butler's swingeing critique of the sex/gender distinction in *Gender Trouble* (1990), which spilled over into the development of gay and queer studies.

It should be noted that Gatens' (1996) critique of gender feminism, which was first published in the early 1980s, is underscored by a concern to maintain a theory of sexual difference. She argues that 'the neutralization of sexual difference' through the insistence on the sex/gender distinction 'lends itself to those groups or individuals whose analyses reveal a desire to ignore sexual difference and prioritize "class", "discourse", "power" or some other "hobby horse" ' (17). In so doing, such accounts 'reduce sexual politics to gender difference' and privilege the relations between 'gender and power, gender and discourse, or gender and class – as if women's bodies and the representation and control of women's *bodies* were not a crucial site in these struggles' (ibid.). That is, such moves take attention away from the centrality of women's bodies as a site for feminist engagement with what she terms, 'imaginary bodies'.

The matter of bodies is also central to Butler's critique of feminist discourses, as indicated in the previous chapter. In *Gender Trouble* (1990), however, she not only challenges feminist constructionism but she also questions feminist foundationalism. Butler is only too aware that the notion of the subject of 'woman' as a stable and fixed entity, in terms of representation and politics, has recently been called into question. Despite the fact that claims of a global patriarchy inherent in earlier radical feminist viewpoints have been discredited, Butler nevertheless recognizes that the idea of 'woman' has been much harder to give up in certain feminist discourse. Butler takes issue with feminist theory that holds on to the idea that there is some kind of pre-existing universal identity, which might be categorized as 'woman'. She suggests that if Foucault's analysis of how the human subject is discursively constituted and regulated by systems of juridical power is correct, it follows that feminism may also be considered as one of the mechanisms of power that produces the subject (woman) that it is supposed to liberate. Feminist critique, she argues, must question and understand how the feminist subject is produced and constrained by the discursive practices of feminism, which seek to liberate it.

Following Foucault, Butler advocates a feminist *genealogy* of the category of 'woman', in order to 'trace the political operations that produce and conceal what qualifies as the juridical subject of feminism' (5). She questions the foundationalism inherent in feminist politics in regard to the formation of the subject. At the same time she challenges the 'compulsory order of sex/gender/desire' or 'heterosexual normativity'. In so doing she critiques the cornerstone of the sex/gender distinction in feminism, which, as discussed earlier, sought to overcome the 'biology as destiny' argument inherent in western thought, but in so doing, ended up by fixing the category of sex, while privileging gender as a free floating cultural signifier. Butler argues that:

> If gender is the cultural meanings that the sexed body assumes, then a gender cannot be said to follow from sex in any one way. Taken to its logical limit, this sex/gender distinction suggests a radical discontinuity between sexed bodies and culturally constructed genders.
>
> (Butler 1990: 6)

Thus, the construction of (gendered) men, for example, does not necessarily follow through to the bodies of males alone, and the same logic applies to the construction of (gendered) women. She also suggests that the presumed binary divide between male and female bodies should not necessarily give rise to binary gender constructions; that is, 'there is no reason to assume that the genders ought to remain as two' (ibid.). But, for Butler, the presumed binary divide within feminist discourse gives rise to another set of issues. How is it possible, she asks, to speak of the 'givenness' of sex and/or gender without in the first place asking how, and under what circumstances sex and/or gender is viewed as given? Further, 'what is "sex" ' in the first instance? When we speak of sex, do we mean that it is natural? If so, what is this naturalness based on – anatomical and/or physiological principles and so on? Following writers like Foucault (1984) and Thomas Laqueur (1987), who point to the notion of 'histories of sexualities', Butler questions the fact that sex is a natural fact of existence, by suggesting that such supposed natural facts are discursively produced within certain political discourses. That is, there is nothing 'natural' about them.

> If the immutable character of sex is contested, perhaps this construct called 'sex' is as culturally constructed as gender; indeed, perhaps it was always

> already gender, with the consequence that the distinction between sex and gender turns out to be no distinction at all.
>
> (Butler 1990: 7)

Butler considers that the positioning of sex in a 'prediscursive' frame, that is, in a domain that lies outside language and discourse (i.e. in the realms of the natural as opposed to culture), facilitates the stability and the fixity of the sexual binary divide, and insists on showing it up. Rather, for Butler, the production of sex as 'prediscursive' needs to be understood as the effect of the apparatus of cultural construction designated by gender. For Butler, 'sex' is an *effect* of the ways in which the normative structure of gender (compulsory heterosexuality) is constructed, produced and sustained. Gender, in Butler's terms, is a 'doing word', an action, and in everyday life we routinely perform gender through a series of repeated bodily performances. 'The effect of gender is produced through the stylization of the body, and hence, must be understood as the mundane way in which bodily gestures, movements and styles of various kinds constitute the illusion of an abiding gendered self' (140).

In a number of respects, this statement resonates with Goffman's and Bourdieu's ideas on habitual bodily practices discussed in the previous chapter and the importance of these to gender formation and social structure. Gender can be seen as an act, which, like 'other ritual social dramas', necessitates that 'the performance is repeated' (ibid.). Here Butler makes reference to Victor Turner's (1974) notion of social drama. She maintains that gender should not be viewed as 'a stable entity or locus of agency, from which various acts follow' (ibid.). Instead, gender needs to be seen as 'an identity tenuously constituted in time, instituted in exterior space through *a stylized repetition of acts*' (ibid.). So gender turns out to be a 'fabrication'. The normative binary sexual divide is secured as 'natural' through a range of discursive practices and the iterative processes of 'performative accomplishments'. Sex, then, according to Butler, is no more 'natural', 'fixed' or stable than the category of 'gender' and as such, they are both open, at least potentially and within limits, to the possibility of resistance and change. But it is important to note that Butler, like Derrida and Foucault, rejects the sociological notion of 'agency', or individual conscious action, in favour of a decentring of the subject.

Contrary to earlier feminist perceptions of 'parodic identities' such as drag and cross-dressing as being degrading to women, Butler considers these kinds of 'performances' as potentially challenging the heterosexual status quo. As indicated in the previous chapter, Butler suggests that the parodying of gender identity in 'cultural practices such as drag, cross-dressing, and the sexual stylizations of butch/femme' identities may reveal gender identity as a 'fabricating mechanism through which social construction takes place' (137). She also envisages a more complex relation between the notion of 'imitation' and the 'original' than the feminist critique of such practices usually imply. A drag performance, for example, she suggests, involves 'three contingent dimensions of significant corporeality: anatomical sex, gender identity, and gender performance' (ibid.). The drag performance, usually a man performing as a woman, plays on the fact that the performer's (male) anatomy is separate from that of the (feminine) gender which is being performed. Moreover, both of these are set apart from the 'gender of the performance'. As a consequence, 'the performance suggests a dissonance not only between sex and performance, but sex and gender, and gender and performance' (ibid.).

> *In imitating gender, drag implicitly reveals the imitative structure of gender itself – as well as its contingency* . . . In the place of the law of heterosexual coherence, we see sex denaturalized by means of a performance which avows their distinctness and dramatizes the cultural mechanism of their fabricated unity.
>
> (Butler 1990: 137–8)

In Butler's framework, it becomes impossible to speak of 'the body' as such; rather, her approach may be seen as an undoing of the normative construction of 'the body' in social and cultural analyses. In a sense, she pushes Foucault's discursive approach to the body, which, as I have argued elsewhere (Thomas 2003), retains a version of the materiality of bodies, to its radical conclusion. In Butler's work, there really is 'no body' other than that which is constructed and produced in and through the effects of heteronormative discursive practices and so, the body becomes a 'text'. While Susan Bordo (1993) recognizes Butler's 'brilliance' at 'detecting and deconstructing naturalist assumptions', she also argues that Butler's approach ends up by recasting 'all biological claims' within a more broadly based apparatus 'that sees discourse as foundation and the body as thoroughly "text" ' (290–91). Although Butler rejects feminist foundationalism with regard to the naturalness of binary sex/gender divide, and deftly shows it up for what it is, it may be argued that her remedy is to replace one type of foundationalism for another; 'a discursive or linguistic foundationalism as the highest critical court, the clarifying, demystifying Truth' (291). Bordo argues that:

> Butler's world is one in which language swallows everything up, voraciously, a theoretical pasta machine through which the categories of competing frameworks are pressed and reprocessed as 'tropes'. . . . Butler's analyses of how gender is constituted or subverted take the body as just such a text whose meanings can be analyzed in abstraction from experience, history, material practice, and context.
>
> (Bordo 1993: 291–2)

Bordo's, it must be said, is explicitly critical of some of the key assumptions in postmodernist theory, whilst recognizing the significant insights it offers. Much like Foucault, she maintains a cultural constructionist position which sees the presentation of biological body as always being culturally and politically 'inscribed' and shaped, and subject to change and resistance. However, she rejects the idea (inherent in Butler's work and that of other postmodern feminist writing), 'that the body is itself a fiction' (228). Nevertheless, Bordo uses the notion of bodily 'inscription' a great deal, which does suggest that the body is 'written on' by cultural forces, although she argues against the idea of the body as a tabula rasa, a blank slate waiting to be moulded by culture. Rather than meaning being given a 'free rein', through notions of difference, multiplicity and subversive textual strategies such as parody, it is necessary, Bordo argues, to situate the 'material body's locatedness in history practice and culture' (38). She suggests that contrary to the postmodernist turn towards agency and voluntarism, where the body is viewed as malleable and subject to change and alteration at will, pervasive ideas and discourses of gender, race or class for that matter, are hard to escape in everyday life. This is despite 'the effacement of the body' which such discourses imply, which in turn might lead us to ask of Butler's radical discursive approach, 'is there *a body* in this text?' (ibid.).

Butler, of course, responded to such criticisms in her subsequent book (1993: 16) by considering 'the matter or bodies as a kind of materialization governed by regulatory norms'. She proposes that it is through discourses that the bodies are materialized and authorized with powers of agency. However, according to Ian Burkitt (1999: 95), this does not entirely answer the criticism of the lack of materiality of bodies in

Butler's work 'because discourse is still idealized as the precondition of all being . . . and the material body itself'.

Bordo (1993) argues that the effacement of the material body evident in textual approaches is also evident in popular and (post)feminist discourses which address important 'gender' issues such as cosmetic surgery, in which the tropes of agency, self-help and even heroism are to be found (see Negrin 2002; Fraser 2003). Earlier feminist criticisms of plastic surgery tended to focus on the idea of woman as a victim of capitalist, patriarchal society. Surgery issues feature quite strongly in the discussion of the two performance case studies that follow, Jo Spence and Orlan, whose own bodies feature strongly in their work. Here, attention is drawn to particular aspects of their work, and what these have to say about gendered bodies and their approach to feminism. As the foregone discussion has been heavily tilted towards theory, the aim, here, is to consider how these theories are worked through in different kinds of performance practices, which involve challenging the construction and production of the representation of women's bodies in contemporary culture.

JO SPENCE – 'PUTTING MYSELF IN THE PICTURE'

Jo Spence's 'political personal photographic autobiography', *Putting Myself in the Picture* (1986) emerged out of an exhibition on her 'involvement with photography from 1950–1986', which included her written work as well as her photography. This book and the posthumous collection published in 1995, three years after her death, testify to the development and range of her challenging photographic practices and principles, her writing and her engagement with leftist politics, feminism and cultural theory. Spence described herself in the aforementioned texts as a 'socialist and a feminist' or sometimes as a 'socialist feminist'. She also referred to herself as a 'photographic worker', a 'cultural worker' and on occasion as an 'artist' (almost always in inverted commas). Although I mostly discuss Spence's 'individual' approach here, it is important to bear in mind that she generally worked in a collaborative manner from around 1974 through, for example, the Photographic Workshop, which she co-founded and the mostly women's collective, the Hackney Flashers.

After spending a number of years as a high street photographer, Spence turned towards a documentary mode of photography in the 1970s, as did a number of leftist oriented cultural workers, community activists and collectives during that period, whose interests were increasingly directed towards identity politics. She later abandoned this in favour of photo montage using the non-naturalistic methods of other media, such as the theatre. From around 1979, Spence's work became more directed towards biographical and autobiographical considerations and events, which in turn led her to develop (with Rosie Martin) 'phototherapy', which means 'using photography to heal ourselves' (Spence 1986).

In the Preface of Spence's book, *Cultural Sniping* (1995), Annette Kuhn points to the different aspects of Spence's work over the years, but she also detects evidence of core themes throughout her work:

> If there is a thread that runs through all of Jo Spence's work, it is surely this capacity to challenge the boundaries between inner and outer, private and public, personal and political; and in so doing 'make strange' the distinctions which pervade our culture and shape the ways we think about ourselves and our lives.
>
> (Spence 1995: 19)

I first came across Spence's writing and photography in the early 1980s when I was lecturing on feminist aesthetics as part of a course on the sociology of art. What struck attention from the outset was the fact that from around the late 1970s she was increasingly putting herself in the picture frame, as it were, and challenging the conventions of the 'to-be-looked-at-ness' of women (Mulvey 1975), as were several women 'artists' at this time (see for example, Parker and Pollock 1987). Spence's 'auto'-photography, for want of a better term, somehow seemed at once to be more 'real', vulnerable, challenging and funny, too, perhaps because photography is such a familiar medium in our everyday lives. Yet, the images of the ordinary, naked or half-dressed female body performing in settings where we would usually expect a clothed body, also have a distancing effect; that is, they make you stand back and reflect on what is going on in the frame. Take, for example, the photo of Spence standing, bare-footed, with head held high, just inside the door of a (working-class?) house (with number 152 on the wall), with two bottles of milk on the ground outside the door. She is holding a long-handled broom in front of her with one hand, and is wearing a skirt tied around her waist, and a strand of beads round her neck on her otherwise bare upper-body; the title of the photo is 'Colonization' (Spence 1986: 124). Projects such as *Remodelling Photo History* (with Terry Dennett 1981–1982) utilized the Brechtian technique of distanciation to 'make strange' the formal aesthetic codes and institutional practices that photographers common-sensically and without question, deploy on a daily basis. This developed into a 'form of photo-theatre', which enabled Spence and Dennett to make use of 'non-naturalistic modes of representation' to generate a form of 'hybrid "spectacle" whilst drawing upon and disrupting well-known genres of photography which have been concerned with the representations of aspects of the female body' (119–120). Exhibitions like *Beyond the Family Album*, which was shown at the Hayward Gallery in 1979, spoke to the emergent crisis of representation discussed in the first chapter. Her role shifted from 'being behind the camera to being in front of it' and in so doing, she became 'an active rather than a passive subject' (83).

My concern in this section, however, is directed towards the work (photographic and textual) Spence produced around *The Picture of Health* project (1982–1985) which toured from 1985 and her reflections of that period. Spence's work in the early 1980s indicates an adherence to a number of the concerns voiced in gender feminism, particularly socialist feminism, through the control of women's bodies by the (male dominated) medical profession, and the iniquities of a class society in regard to matters of health and illness. At the same time, it telegraphs her move towards psychoanalytic theory and an interest in subjectivity, which, as indicated earlier, were also important influences in the fragmentation of feminist theory.

In a conversation between Spence and the feminist writer Ros Coward, published in 1986, the former expressed 'increasing frustration at reading more and more things that are coming into print about the body from an academic or theoretical point of view' (Coward and Spence 1986: 24). Spence, here and elsewhere (1995), adopts a sceptical view towards the notion of social constructionism, although, as we shall see, her approach does contain a strong sense of the potential for agency through consciousness and practice. While she recognizes that we experience the body 'in the imagination as a social construct', she is very clear that this is it not all there is. In the first chapter, I noted that in everyday life we do not often think very much about the inner workings of our bodies until something forces us to. Spence became acutely aware of the workings of her own body, of which she came to realize she knew very little, when she was diagnosed with breast cancer. This experience confirmed her in the view that the body is not simply a social construct, and that she had been operating on the basis of a mind/body separation:

> . . . having had a breast labelled as cancerous in 1982 rapidly changes the way I viewed my body . . . if the body is deteriorating in the organic, cellular sense, then it is nonsense to talk about social constructs or the imaginary.
>
> (Coward and Spence 1986: 24)

Thus, for Spence, the body has real substance. It is not an 'image, or idea or psychic structure'; it is not simply the effect of discourse, rather it is 'made up of blood, bones and tissue' (Spence 1986: 151).

The shock of cancer also convinced Spence that she had to learn more about her body, her health and her illness, in order to fight it. Her experience at the hands of the medical practitioners in the National Health Service (NHS) hospital, whose care she was under initially, made her determined to wrest control of her body from the hands of orthodox medicine and find an alternative route that would treat the whole person. Orthodox medicine, Spence argued, viewed the body as individual parts not as a whole; it treated the symptoms, not the underlying causes of illness and disease. Despite the consultant's advice that she should have her left breast removed, along with the view embedded in much medical discourse that not to do as the consultant advised in this situation was tantamount to committing 'suicide', and her own terror, Spence somehow found the voice to refuse to go down this line. She insisted on having only the cancerous growth removed and she also refused radiotherapy and chemotherapy.

Spence embarked on a photographic project to document her journey through the operative process and the alternative route she eventually chose to take in regard to managing her illness, her health and well-being, which she found in traditional Chinese medicine. At the same time, she came to recognize that she had to find a different voice from her usual socialist feminist confrontational stance, one that would facilitate a care of the self from a holistic perspective, if she were to survive. In this way, she began a 'research project on the politics of cancer' (151), charting her pathway through the medical process within orthodox medicine and her journey out of it. The aim, according to Spence, was 'to use the photos as a starting point of the stories' she wanted to tell about this experience and these photographs offered her 'an entry point' (Coward and Spence 1986: 35). There are several interconnecting strands that run through these story lines, which relate to her continuing interest in challenging the boundaries 'between inner and outer, private and public, and personal and political', as indicated earlier.

Spence had a friend photograph her going into the consultant's room to be given the results of her tests. When she went into hospital she took with her a tableau of herself with 'THE PROPERTY OF JO SPENCE?' written on her left breast as a reminder that she had rights over her body (157). From her bed, she photographed the consultant's ward rounds (158) and the images she encountered as she was being wheeled to the operating theatre after the pre-med. After the operation, the ward doctor photographed her being wheeled out of the operation into the recovery room (159–60).

Although Spence photographed her operation from a patient's viewpoint, lying on her back, the style was not necessarily autobiographical or in documentary mode in the direct sense. For example, the photographs taken of the surgeon and nurses in the operating theatre (171) were fashioned 'in a style of a doctor/nurse photo-romance in order to make a rift in our perception of what we experience (but can't talk about), and previous ideologies which we carry with us about the doctor as a romantic hero' (Coward and Spence 1986: 38). The concern, here, was not only to 'problematise visual representation', but also to 'de-neutralise orthodox medicine itself, so that we can begin to ask questions about it as women (people)' (ibid.).

In order to overcome the alienation she felt in regard to her body, Spence needed to find a way of taking 'responsibility' for what was happening to her. This was not simply an act of individual agency. Rather, it can be seen as an act of ethical responsibility to undo the damage to her body that in large part, she believed, had come about because of the way in which she was socialized within her (working) class to 'neglect' herself 'materially, environmentally, economically and . . . spiritually' (Spence 1986: 152). Spence recognized that she had also been collusive in this and so, there was an element of agency or relative autonomy at work in this.

In retrospect, Spence came to see that the 'journey' she was taken on as a consequence of being diagnosed with breast cancer led her to develop a different, more harmonious relation to her body and a 'new practice in photography', which would eventually converge in to phototherapy. This involved harnessing the tools of psychoanalytic theory to the practice of photography for understanding her own past, and for the many others like her, through amateur and family photography (1995: 149). This was in contrast to the overwhelming preference for submitting mass-produced or public images to 'the endless deconstruction of texts', which emerged out of debates in film theory in the early 1980s. The importance of family photography, for Spence, stemmed from her abiding interest in 'the ways in which, in a patriarchal and class-based society, family relationships are contradictory and unevenly lived out through the grind of fantasy and experience' (150). Spence was aware that the genre of the family photograph is frequently employed in 'advertising photography' to work on the ways 'our unconscious processes can be engaged with'. 'Why not', she asked, 're-appropriate' this from the advertisers, 'turn [it] on its head and deconstruct their techniques back into the family itself?' (ibid.) Spence's background adherence to class and gender and her shift towards subjectivity and a more complex and perhaps subtle, analysis of the relation of the self in society, are clearly recognizable in this context.

Spence was only too well aware of the problem of the silence around breast cancer at that point in time and the lack of clear information on the extent of the problem. She did not keep her 'useful social knowledge' to herself, but went on the campaign trail, writing and lecturing on the facts of cancer and offering the practical solutions that she had learned from the life-changing experience through which she came to realize how much the body was 'not just a bloody sign'. Her challenging autobiographical approach to dealing with breast cancer followed in the wake of the groundbreaking work of other arts practitioners such as black feminist writer Audre Lorde (1980, 1988) and foreshadowed more recent work in the arena of photography such as that of Matuschka, the artist and former model (see Amaya 2004 on this).

Spence recognized the import of the female breast in the sexual cultural imaginary and in women's lived experiences. She challenged the notion that a woman who has a breast removed is no longer a 'real' woman, to others or herself. As far as I am aware, Spence does not address plastic or reconstructive surgery directly in her work. At the time of her diagnosis, this option was not so readily available in the UK on the NHS to women after a mastectomy. However, she does question why women should be 'pressurised' into wearing a prosthesis after this operation on the grounds 'that it has no function other than a *cosmetic* one' (1995: 125). One of Spence's photographs offers us an image of a breast prosthesis resting on a plate, whilst another has a false breast lying on top of a magazine entitled *Medeconomics*, with a tag beside it with the words, 'Mrs Smith Thursday 4 pm', written on it (1986: 171); the former perhaps, conjuring up a link between food, women and sexuality and the latter a connection between medicine and economics and the consequences for women's health.

> Just as social and political injustices are covered up, so (with the 'honourable' exception of war veterans) are injuries, deformities and amputations hidden. This not only leaves the individual isolated, unable to share experiences with other survivors, it also conceals the high incidence of the injury. The tyranny of the one acceptable body shape is maintained.
>
> (Spence 1995: 125)

It follows from this that breast cancer survivors should be encouraged to wear their scars with pride, instead of trying to hide them and 'pass' as 'normal' or 'whole'. Sander Gilman (1999), in his illuminating study on the history of 'aesthetic' surgery (generally referred to as cosmetic surgery in the UK), argues that the idea of 'passing', of merging into the crowd, has been central to its historical development. Although the wearing of a prosthesis may erase the loss of the breast from the outside, in terms of being looked at by others, Spence recognized that the physical (and psychic) scars that women experience are not so easy to erase, with the result that women are on hyper-alert in case the disjuncture between the outer appearing body and the real experiential body is revealed. One possible pay-off for not accepting the prosthesis is that the notion of 'different bodies' may come to be seen in a positive light, if enough women 'survivors' choose this route. Yet, Spence stated that she 'did not want to be *mutilated* beyond the three vivid slashes' that remained after the lumpectomy (1986: 162 my emphasis) and she recalled the terror of realizing that she could lose one or even two breasts to the illness. By positing the notion of the breast prosthesis and presumably therefore, breast reconstruction, as purely 'cosmetic', meaning that it is unnecessary, Spence reveals her attachment to what is largely a second wave feminist viewpoint on this matter. However, as Gilman's (1999) study demonstrates, the line between reconstructive surgery and cosmetic surgery is blurred round the edges.

The debates in America in the 1990s around breast implants, which arose as a consequence of the health scares associated with silicone gel implants, often centred on the need or otherwise for such a procedure. In 1992, the Commissioner for the US Food and Drug Administration Board 'banned the use of silicone gel implants for aesthetic reasons, but stated that they could be used for post-mastectomy reconstruction' (Gilman 1999: 243), despite the fact that it was accepted by the medical establishment that such implants were a danger to women's health. The rationale behind this, according to Gilman, was that 'the danger of the post-mastectomy patient's unhappiness with her now scarred body was greater than the physical risk from the implant' (ibid.). The implication is that the silicone implant will aid the recovery of the patient and perhaps, would allow her to 'pass', despite the dangers associated the implant bursting and causing further serious harm. This implies a disjuncture between aesthetic and reconstructive surgery, with the former being treated as having less medical value and inhering notions of vanity. In turn, this sets up a distinction between women who 'need' breast implants and women who merely 'want' breast augmentation. The 'reconstruction of the erotic female body', according to Gilman, was a primary aim of reconstructive surgery. However, as discussed above, writers such as Lorde (1980) and artists like Matuschka (see Schlessinger 1997; http://www.matuschka.net/old/interviews/gtgoffchest.html), who had a breast removed as a result of cancer, have argued that the asymmetric body of a woman who has had a mastectomy should be treated as an acceptable and erotic body. Perhaps, then, Spence's 'survivor' perspective was not so far off the mark after all?

Jo Spence developed her artistic practice, using her own everyday body as both subject and object of her work, as a consequence of her experiences of having cancer and having to go under the surgeon's knife. The performance artist, Orlan, on the other hand, who claims that there is no distinction between her body and her art, chose to go under the surgeon's knife in a series of cosmetic surgical operations, as part and parcel of her performance practice in the early 1990s. Indeed, Orlan's actual surgical operations were the centre pieces of her critical performance practice; the operating theatre was *the* performance. In a number of respects, Orlan's claims to feminism reside within the general frame of postmodernism. In certain respects, however, her work may be seen to challenge the very ground of the voluntarist imperative within postmodern discourse. She can also be seen to critique and support the discourse of modern cosmetic surgery, a practice that has grown exponentially in the UK in recent years. The following section explores these issues in the context of Orlan's work and her approach to the body.

ORLAN: 'I HAVE GIVEN MY BODY TO ART'

Just as Jo Spence embarked on 'politics of cancer' project in the 1980s, the French multi-media performance artist, Orlan, began a project a decade later entitled, 'The Reincarnation of Orlan', which, as indicated above, consisted of performances of cosmetic surgery. The question of identity has been central to Orlan's practice throughout her four-decade career, as indeed it was to that of Spence, although in a rather different way. Unlike Spence, Orlan (born in St. Etienne, France in 1947) deliberately shied away from revealing too much about the details of her origins and upbringing for much of her career, preferring to keep them vague and thus, it could be implied, intriguing. The result is that we are kept guessing and are never quite sure who Orlan actually *is*. The deliberate haziness surrounding her background, according the art critic Barbara Rose, was useful 'to maintain the anonymity required to project an enigmatic "star quality" ' (1993: 84). However, since Orlan gained international star status from her surgical operations, she appears to be more willing to reveal the facts of her biography, which are publicly available on her official website (http://www.orlan.net/). Nevertheless, the question of identity is key to her work, pre and post operation-performances.

Orlan's name, as several commentators have noted (Wilson 1996; Ince 2000) gives us a glimpse of the slipperiness of the artist's identity and her unwillingness to be fixed. The name Orlan is an adopted (stage) name and according to Kate Ince, 'the polysemic and cultural connotations' of the name itself is but one example 'of how pivotal the question of identity has been in Orlan's career' (Ince 2000: 1). For example, it sounds like a man's name. It has connotations with Virginia Woolf's hero(ine) *Orlando*, who changed sex back and forth over generations. It could refer to the name of the perfume Orlane. The O could refer to 'Other' or the O in the pornographic book, *The Story of O*, 'the O that signifies and figures in the opening of all orifices' (ibid.). It could also refer to St. Joan of Arc, who was the Maid of *Orléans*, and as indicated below, Orlan added saintliness to her performance persona in the 1970s.

Orlan's passage 'from the art gallery to the operating room' began in the 1960s, when she 'improvised her first performances and public spectacles' during the period of the liberation movements, which culminated in the 1968 May events in France (Rose 1993: 83). Like many artists in France and the US in the 1960s, Orlan was influenced by the early twentieth century dadaist (anti-art) approach of Marcel Duchamp, who developed the notion of the 'readymade' (non)art work, via the avant-garde artists she came across in that period (ibid.). The postmodern dancer, Yvonne Rainer, whose postmodern approach to dance was discussed in the previous chapter, took up the notion of the everyday, readymade movement as the basis for her performance practice. Orlan's (1996) take on this, as discussed below, was to treat her own body as a readymade. As a performance artist, Orlan's body has been the central medium of expression of her artistic practice throughout her career. Indeed, she maintains that her body and her art are indivisible: 'My body is my art.' She identifies her work as 'carnal art', in order to distinguish her approach from 'corporeal art' (88).

In 1971, Orlan adopted the title of Saint Orlan for her performance persona. Much of her performance and installation work during the 1970s and 1980s was concerned with religious iconography, particularly Catholic imagery. The performances and installations, which were often shown in churches, drew on the 'conflict-ridden store of art history in which sexuality and art history coincide', using the tropes of 'ornate pastiche' or 'pointedly disrespectful parody' (Ince 2000: 2), which showed up the hypocritical manner in which the female image has been 'split between virgin and whore' (Rose 1993). Orlan's challenges to art history's ideas about beauty and women in religious imagery are also evident in her surgical performances of the 1990s. In addition to critiquing the language of art and religion, and being informed by and critical of the universalist aspects of psychoanalytic theory, Orlan's project is also feminist:

> . . . Orlan has worked with the relationship between her own image and that of Baroque religious iconography, using historical references to relate contemporary artistic practice to religious imagery. She has also staged performances which question the relationships between the institutional structures of art and religion, and the roles which have been relegated to women within these institutions.
>
> (Wilson 1996: 5)

Orlan has also insisted that her work 'has always interrogated the status of the feminine body . . . those of the present or past' and that the 'variety of images' she has produced has been concerned with 'the problem of identity and variety' (1996: 84). Despite her feminist stance, Orlan has demonstrated against particular French feminist positions that are founded on the notion of 'woman' as singular essence. For example in the late 1960s, she disrupted several feminist meetings by carrying a placard stating, 'I am a man (un homme) and a woman (une femme)' (85) and in so doing, questioned the notion of the binary sexual divide.

Although Orlan was recognized in France as an avant-garde artist in the late 1980s, her interventionist feminist challenge to the arts establishment in France only received international critical attention and notoriety in the 1990s, as a consequence of the art and popular media hype and controversy surrounding her surgical performances. Part of the reason for this, according to Ince (2000), lies in the fact that while French women's writing, as discussed earlier, attracted much attention in Anglo-American cultural thought in general and feminism in particular in the 1980s, this was not the case with the work of French women visual artists, which had generally received a less than positive response in France from the late 1960s. Thus, as Ince remarks, France failed to take account of the theoretical and critical insights that 'she herself had produced and inspired' in Anglo-American cultural thought (6). While Anglo-American performance was well documented, French performance art from the 1960s was not. All of this, however, was about to change with the Reincarnation of Saint Orlan project, at the end of which the artist was to be renamed.

Carnal Art (Art Charnel), the first in the series of planned and well-documented operation-performances (she had nine in total), which was to take Orlan on the road to 'surgical self re-invention', took place in Paris in July 1990. It was in this first operation-performance that the operating theatre became her studio (Orlan 1996). Orlan had liposuction to her thighs and her face and six days later she had an implant inserted into her chin (modelled on Botticelli's Venus). Orlan's 'explicit body in performance' resonates with that of other performance artists of the period in Europe and the US, such as Annie Sprinkle (see Schneider 1997). Each operation aimed to focus on a specific part of her face. In operations seven, eight and nine, for example, Orlan (1996: 90) insisted on having the largest implants her facial anatomy could take; including two silicone implants that are 'usually to enhance the cheeks', inserted into her temples 'to create two bumps', despite the then public debates over the safety of breast implants. Orlan's stated concern was not to re-invent herself by moulding and lifting her face to improve her 'original' face and give it a more 'natural' or youthful appearance, which is often the rationale embedded in the rhetoric of modern cosmetic surgery, a practice which is overwhelmingly targeted at women (Ince 2000; Fraser 2003). Rather, she intended to use her body as a 'medium of transformation' (Rose 1993). Orlan viewed the operation-performances as 'interventions'.

Orlan had each aspect of the face surgically remodelled to correspond to 'a specific feature of a different great icon in the history of Western art' (Ince 2000: 6). She literally now has the nose of Diana taken from a sculpture from the School of Fontainebleau: 'the mouth of Boucher's Europa, the chin of Botticelli's Venus, the forehead of Leonardo's Da Vinci's Mona Lisa and the eyes of Gérard's Psyche' (ibid.). The choice of these icons was not based on the 'the canons of beauty they are supposed to represent but for their histories' (Orlan 1996: 88): Diana was chosen because she was 'aggressive' and 'insubordinate to the gods' and Mona Lisa was selected because she is not 'beautiful according to present standards of beauty' and because of her androgynous look (a self-portrait of Da Vinci lies beneath the surface of the painting). Venus was chosen for her association with fertility and creativity, Europa was selected because the picture is unfinished and her look is directed towards another continent and thus, an unknown future, and Psyche was favoured because of her desire for 'love and spiritual beauty'. Orlan's composite face, according to Ince, deconstructs the ideal of a 'feminine aesthetic of unity'. It can therefore be seen as an attempt to undercut 'any theory of female subjectivity that entertains a universalist, phallogocentric concept of the Woman' (Ince 2000: 123).

Orlan's nine operation-performances took place between 1990 and 1993. They were highly theatrical events in which her flesh, not just her outer appearing body, was the new medium of expression, shown and documented in all its graphic detail (Rose 1993). Indeed, it could be argued that the operations were concerned to undo the distinction between the outer and inner body. These performances were highly choreographed and directed by Orlan and each had its own style. All of the operations in the Reincarnation project employed 'the use of quasi theatrical sets . . . while also offering representational complexity and a degree of staginess that borders on camp', particularly the seventh operation entitled Omnipresence (Ince 2000: 7). The operation-performances involved 'interactive communication, often with an international audience via fax and live satellite, music, dance and elaborate costumes' (Hirschhorn 1996: 4). For instance, the surgical team and Orlan's assistants wore clothing by fashion designers like Paco Rabanne and Issey Miyake.

The operation-performances mixed theatricality of the theatre proper with the drama of the operating theatre, with all of its 'blood and gore', which is part of the fascination of televisual medical documentaries and hospital drama series (Ince 2000). During the operations, Orlan remained completely conscious (except for one operation), having had an epidural injection in her back (itself a very risky business) and local anaesthetics to mask the pain. While her flesh was being opened and re-sculpted, Orlan read from psychoanalytic and literary texts such as Kristeva's *Powers of Horror* (1982), the Lacanian inspired writings of Ugénie Lemoine-Luccuoni, and the work of Antonin Artaud, the writer, actor and theatre director, who proposed a 'theatre of cruelty', which Orlan's operation-performances seemed to metaphorically and literally embody.

Orlan (1996: 91) was 'the first artist to use surgery as a medium and to alter the purpose of cosmetic surgery: to look better, to look young'. She considers that cosmetic surgery 'is one of the areas in which man's power over the body of a woman can inscribe itself most strongly' (ibid.). Nevertheless, and perhaps unsurprisingly, feminists have often accused her of promoting cosmetic surgery (Ince 2000). In part, this is because Orlan (1996) clearly states that she is not against cosmetic surgery per se. Rather, her intention is to take a stand against the ideology of a narrow, standardized notion of beauty which is increasingly imposed on 'the flesh' of women ('and men'). She adopts a postmodern, agenic approach to re-forming the body through cosmetic surgery, which, she maintains, can be very beneficial to some people if they no longer like what they see in the mirror; that is, where their outer appearance does not correspond to their inner image of their self. Cosmetic surgery, her argument implies, can be a way for women to gain control over their bodies by appropriating advances in technology for their own ends.

In her discussion of Orlan's operation-performances, Kathy Davis (2003: 106) notes that there appears to be several continuities between her concern to provide a feminist critique of 'the technologies and practices of the feminist beauty system while taking women who have cosmetic surgery seriously' and Orlan's surgical operation-performances. For example, both Orlan and the women in Davis' study of cosmetic surgery stated that they were not interested in becoming more beautiful, nor did they feel that they were being coerced into surgery by 'social and ideological forces'. Rather, the women, like Orlan, viewed themselves as active agents in the world. In remoulding their bodies they considered that they were also potentially remoulding and reshaping their lives.

However, for Davis, this is where the similarities end; Orlan's concern, Davis states, is not directed towards a 'real-life problem; it is about art' (110). Orlan's quest, Davis argues, is separate from and not reducible to, everyday life. Her body is a means to her 'art, and her personal feelings are entirely irrelevant'. She does not undergo surgery to 'alleviate suffering with her body', but to make a 'public and highly abstract statement about beauty, identity and agency'. When asked about her experiences of pain, Davis points out that Orlan 'merely shrugs and says: "Art is a dirty job, but someone has to do it" ' (111). All of this is very much at odds with the reasons the women in Davis' study give for undergoing cosmetic surgery; their project is both 'private and personal' and they are concerned to remove the 'suffering' they have experienced. They are worried about the pain and the potential pitfalls if surgery goes wrong and they do not seek to publicize their operations, preferring to keep quiet (in order to pass, as Gilman 1999 might argue, perhaps?). Although they touch on issues like 'beauty, identity and agency . . ., they are always linked to their experiences and their particular life histories. Their justification in having cosmetic surgery is necessity . . . They do not care about changing the world; they simply want to change themselves' (ibid.).

Art and life, for Davis, remain different propositions. However, she considers that Orlan's surgical operations may be seen in terms of a feminist utopia, where, as discussed earlier, technology has been embraced by women 'for their own ends'. It should be clear that the status of the body in society in relation to new technologies is central to Orlan's quest to remould and transform her identity, which the artist describes as 'woman-woman transsexualism'. A feminist utopian response, according to Davis, appeals to the imagination on several levels. It not only enables feminists to 'take a stand' against the beauty myth imposed on women, it also reveals the 'excesses of the technological fix'. It renders unstable preconceptions of 'beauty, identity and the female body' and it offers the possibility of 'how women might engage with their bodies in empowering ways' (113). More importantly however, 'it promises the best of both worlds: a chance to be critical of the victimization of women without having to be victims ourselves' (ibid.). Nevertheless, Davis is uneasy about this neat solution and concludes that a utopian response to cosmetic surgery in general and Orlan's practice in particular, 'discounts the suffering' and the potential risk that women routinely go through in relation to this, whilst overstating the 'possibilities of modern technologies' (114). It leaves women who 'suffer' with their appearance and want to change it through surgery appearing as 'cultural dopes', who dance to the tune of the beauty myth. What is left out, in other words, is the ordinary 'sentient embodied female subject, the one who feels concern about herself and about others' (115).

However, Davis can also be criticized for treating women who undergo cosmetic surgery as martyrs (Fraser 2003; Jones 2008). *The Reincarnation of Saint ORLAN*, from the artist's point of view, however, 'is not a story of a martyr, but of a character that dissolves through added multiplicity' (Orlan 2010: 111). Moreover, in regard to cosmetic surgery, we may be witnessing a case of 'life imitating art', if the extraordinary rise and growing acceptance of surgical and non-surgical cosmetic procedures in the UK in recent years, discussed earlier, is of any significance. The women in Davis' study also sought to merge into the crowd. However, there appears to be a shift in this as 'breast enhancement', the fastest growing cosmetic surgical procedure in the UK, and the top surgical procedure in the US since 2008, seems to be more about 'showing' than 'passing' and in this again, perhaps, Orlan was in the advanced guard, but once more in a more reflexive, critical manner.

Orlan was the first artist to undergo cosmetic surgery in the public arena of art, then 'documentary' programmes like Channel Five's *Plastic Surgery Live* and MTV's *I Want a Famous Face*, which utilize 'ordinary' citizens as the performers, surely followed suit, although with considerably less critical reflection (see Tait 2007 on the media aspect of cosmetic surgery). A recent survey for the ASAPS found that more than half the population in the US (51 per cent), regardless of income, were in favour of plastic surgery (ASAPS 2010; http://www.surgery.org/media/news-releases/survey-shows-that-more-than-half-of-americans-approve-of-cosmetic-plastic-surgery). Perhaps Orlan, like Spence, had her finger on the pulse of things to come after all.

In a more positive vein of cultural criticism, Orlan's surgical operation practice has also been viewed 'as a striking instance of Butler's theory of the continual displacement of norms of identity via (re)materialization' and of gender performativity (Ince 2000: 113). Butler's discussion of gender performativity in *Gender Trouble* (1990), as noted earlier, inadvertently led to an outpouring of writing in which gender was treated as something we can just undo and remake at will, rather like putting on a new set of clothes (Butler 1994). In *Bodies that Matter* (1993), Butler sought to rectify this by distinguishing between performance, which supposes a voluntarist notion of subjectivity on the one hand and performativity on the other, which confronts the very idea of the subject. According to Ince, Orlan's surgical performances may be seen to speak to both performance and performativity at the same time; 'she performs performativity, whilst performativity performs her' (2000: 113). Thus, I would suggest, breaking down the perceived distinctions between performance in everyday life and artistic practice, as discussed before.

CONCLUSION

This chapter began by examining two reference points for theorizing the gendered body, second wave feminism and postmodern feminism. Jo Spence's photographic practice and Orlan's performance practice were used to explore these positions, while recognizing similarities and differences between and within these feminist viewpoints. While Jo Spence's approach to the body may be seen to be situated largely within a second wave socialist feminist approach, it is clear that Orlan's approach may be viewed in terms of a postmodernist feminist viewpoint, with more than a hint of utopian feminism. Orlan's *Reincarnation* project 'represents the postmodern celebration of identity as fragmented, multiple, and – above all – fluctuating' (Davis 2003: 109). At the same time, it can be argued that Orlan takes the postmodern textual, anti-biological approach to the body further down the road towards a posthumanist position, as exemplified by the work of Deleuze and Guattari's (1983) development of the notion of 'body without organs'. Like the Australian artist Stelarc, Orlan (1996) considers that the body is becoming 'obsolete' and 'is no longer adequate for the situation'. With the advances of modern technologies, our bodies will become a 'costume' or a 'vehicle' to be transformed in our pursuit of becoming who we are, or wish to be (Davis 2003: 108). However, it remains to be seen just how far this returns us to the much criticized notion of the 'body as machine' embedded in mind/body dichotomy entrenched in the western cultural tradition, which both feminists and postmodernists have systematically tried to overcome. The discussion of Spence's work points to a shift towards a more complex notion of subjectivity, which leaves her about to fall off the edge of her second wave socialist feminist platform. Similarly, Orlan's seemingly postmodernist operation-performances invoke the possibility of the posthuman body, although, as with most of Orlan's work, there are critical implications here too, which are suggested by her later digitized, self-transformation art work, *Self-Hybridizations* (Orlan 2010).

4

ETHNOGRAPHIES OF THE BODY: ABSENT-PRESENCE

INTRODUCTION

The 'body' focus of this chapter is centred on ethnographic research. A brief survey of ethnography would reveal that bodies and bodily practices are scattered here and there across the historical ethnographic landscape, in social anthropology in particular, where they have been studied cross-culturally as well as theorized at various points in the tradition (Kroeber 1952; Mauss 1973 [1935]; Douglas 1970; Polhemus 1978; Csordas 1993, 2002). Bodies have come into visibility in some sub-cultural studies through the focus on the dress, demeanour and argot of post-Second World War urban youth (Bennett 2000; Muggleton 2000) and the moral panics associated with their dance and music cultures like rave (Bennett 2000; Malbon 1999). Studies of the relations between space and place (Goffman 1972; Pile and Thrift 1995; Nast and Pile 1998) have also brought bodies into visibility. As indicated before, although debates around the emergent sociology of the body in the 1990s privileged theory over empirical analysis, a strong body of work has now emerged from a range of sub-disciplinary areas which takes a more 'grounded' approach to embodied practices (Cunningham-Burley and Backett-Milburn 2001; Watson and Cunningham-Burley 2001). Even theoretical 'body' sociologists such as Bryan Turner (see B. Turner and Wainwright 2003), Shilling (2007) and Crossley (2006), have been recently conducting empirical research in this arena.

Despite the fact that bodies and bodily practices have been visible, to a certain extent, across the spectrum of ethnographic studies, there has been a relative lack of attention paid to the embodied character of ethnographic research itself until quite recently (Hastrup 1995; Coffey 1999; Okely 2007). The very idea of observation is an embodied act of and from, the body. While the sociologist/ethnographer may not be 'present' in the text in the form of an 'I', he/she is always implicated in the text through the abstracted authorial voice, if not through his/her acknowledged visceral presence. Even when the ethnographer's self has been addressed explicitly in the text, in relation to gender, age and race, 'the bodily experience of the fieldworker as research process and source of knowledge has been under-scrutinised' (Okely 2007: 66).

This chapter explores a range of viewpoints (traditional and contemporary) concerning ethnography as a mode of inquiry, the methods it incorporates, its shifting character and narrative conventions. The concern is not to consider what ethnography is in a definitive and restrictive manner, but rather to explore how ethnography or, more broadly speaking, qualitative research, has been accomplished. The aim is not to offer a 'how to' approach to the study of the body from an ethnographic perspective. Further, it is important not to be overly prescriptive about what exactly ethnography is, because, as will become evident, there is no one definition of ethnography.

Ethnography as a practice does not 'belong' to one social science discipline and the construction of the history of ethnography is very much rooted (and routed) within disciplinary frameworks. It is not my intention to provide an open and closed historical periodic approach to ethnographic or qualitative research. This periodic approach is perhaps best exemplified by Denzin and Lincoln in the *Handbook of Qualitative Research* (1994), where they set out what they envisage as the key characteristics of the five stages or 'moments' in ethnographic research.

According to Denzin and Lincoln, the first moment in ethnographic research, from 1900 to the Second World War, was objectivist and positivist in orientation and was characterized by the notion of the lone researcher or fieldworker bravely entering into unfamiliar territories. The second 'modernist' moment, from 1945 to the late 1960s, was characterized by a concern to set out the (methodological) tools of the ethnographic trade. This period witnessed a raft of 'rigorous qualitative studies of important social processes including deviance and social control in the classroom and society' (8). The third moment, 'blurred genres', post 1968 to 1986, witnessed the blossoming of a range of theoretical interests and perspectives, along with a variety of 'ways of collecting and analyzing materials' (9). The 'crisis of representation', which was discussed in the first chapter, coincided with the successive fourth moment in ethnography. This was characterized by the erosion of classical traditions of thought, whereby the very truths and fixities on which ethnography was founded were revealed as textual fictions. The authorial voice of the ethnographer, the teller of the ethnographic tale, was called into question and in a kind of role reversal, the observer (ethnographer) became the observed (Stocking 1983), opening up the possibility of a more challenging self-critical approach on behalf of the observer.

The 1990s witnessed the entry into the fifth moment in which different ethnographic sensibilities lived in tension with each other; the broadly 'interpretive, postmodern, feminist and critical sensibility' on the one hand and 'the more narrowly defined, positivist, post-positivist, humanistic and naturalistic conceptions of human experience', on the other (Lincoln and Denzin 1994: 576). In the late 1990s, Denzin (1997) anticipated the possibility that ethnography was moving towards a sixth moment, which is reflexive, 'deeply ethical, open ended, and conflictual, performance and audience based, and always personal, biographical, political, structural, and historical' (266).

Note that there is a rapid succession of 'moments' in the second half of the twentieth century compared with the first half. This is not so surprising given that ethnography as a method was only developed in the first half of the century. However, it may be suggested that whilst the authors point to a diversity of perspectives and modes of analyses in ethnography, particularly in the contemporary era of 'messy texts' (Marcus 1998), they tend to paper over differences and continuities within and across the perceived six developmental moments in ethnography's canon.

This chapter does briefly point to the different histories of ethnography in anthropology and sociology. However, the concern is to direct attention more towards the employment of rhetorical devices and tools which give ethnographic accounts their particular flavour, as in the framework put forward by another social scientist, John Van Maanen (1988), as opposed to the periodic typology offered by Denzin and Lincoln. Once again, and rather inevitably, feminist interventions in social research are never far away from the surface of this discussion, although I will not dwell on these, here, in any detail because many of the issues have already been raised in earlier chapters. The 'crises of representation' in ethnography in the late 1980s is generally assumed to have been given voice through two major publications which heralded the establishment of and debates around, postmodernist ethnography: James Clifford and George Marcus's edited collection, *Writing Culture* (1986) and Marcus and Mike Fischer's volume, *Anthropology as Cultural Critique* (1986). It is worth noting that feminist ethnographers had been challenging the very grounds on which ethnography had been founded for a number of years before these publications burst onto the scene. Nevertheless, their critiques of objectivity and traditional masculinist, patriarchal tropes embedded in ethnographic orthodoxy, were not well received within the mainstream of the field (Caplan 1988; Bell *et al.* 1993, Woodward, Kath 2008) at the time. It is notable that Clifford and Marcus's edition did not include a feminist voice. Although Clifford (1986) noted the importance of feminism's contribution to the issue of gender in ethnography in the introduction to the collection, he also indicated that feminist ethnographers had not contributed 'innovative textual strategies' into the domain. Marcus (1994) later noted the contribution of feminist positioning in ethnography to the development of ethnographic reflexivity.

As with the previous chapter, a case study approach will be taken to illustrate key points. As well as dwelling on specific historical cases, and pointing to the absent-present body in the texts (Leder 1990; Crossley 2006), the latter part of the chapter will look at a particular performative pugilistic bodily practice; boxing, which as it turns out, has much in common with the aesthetically non-pugilistic performative practice of dancing than one might expect at first glance. The 'boxing' case study discussion will be organized around Loïc Wacquant's (2004) visceral first-hand account of a three and a half year immersion study of the 'Sweet science' of boxing in a Chicago gym, along with reference to other studies and writing on boxing. Wacquant worked closely with Bourdieu (Bourdieu and Wacquant 1992) and is generally considered to be the heir to Bourdieu's sociology. Again, the concern is not to detail *what* Wacquant says, rather it is to show *how* he says it and the issues that underpin it. The aim is to consider how studies such as this, which focus on the sentient matter of bodies through intensive participatory embodied practice, impact on the quality of ethnographic and/or qualitative research, as well as the potential pitfalls to which they give rise.

DOING ETHNOGRAPHY – UNDERSTANDING CULTURE

'Ethnography' may be described as a 'written representation of a culture (or selected parts of it)' (Van Maanen 1988: 1). Ethnography's 'trick' is to show the culture or part of it in such a manner that is significant to the reader, 'without great distortion' (13). In this sense, ethnography is conceived of as an end product (a monograph or article), which involves literary or rhetorical features and conventions to create the desired representation of the culture in question, as in, for example, a 'realist' text. The term ethnography also refers to 'a research style' (Atkinson 1990), the doing of ethnography, which generally involves the ethnographer immersing him/herself in the culture or group (the field) under investigation over a lengthy period of time, which is usually longer for the anthropologist than the sociologist. In this sense ethnography refers to 'a particular method or set of methods'. 'In its most characteristic form it involves the ethnographer participating, overtly or covertly, in peoples' daily lives for an extended period of time, watching what happens, listening to what is said, asking questions' (Hammersley and Atkinson 1995: 1).

The core method of ethnography, then, is 'fieldwork' or 'participant observation', and it is essentially qualitative in orientation. Indeed, some would argue that participant observation is the best way to achieve an understanding of others (Stacey 1988), whilst others consider that 'ethnography is guided by an incoherent conception of its own goals' (Hammersley 1992: 11). Ethnography has historically been the favoured approach adopted by social and cultural anthropologists. However, at least to some extent, it could be argued that *all* social researchers are participant observers, which, as Hammersley and Atkinson suggest, means that 'the boundaries around ethnography are necessarily unclear' (1995: 1–2). Ethnographers do not just 'hang out', they also ask questions; a mode of analysis that is usually more closely associated with qualitative social researchers. They also record what people say and do, construct detailed field notes on the daily goings on and interactions in the social settings they are studying and they reflect on these. These features, and more besides, are visible in Wacquant's (2004) visceral ethnography of boxing, which resides within a reflexive sociological framework.

Hybrid interdisciplinary areas of study such as cultural studies, which emerged in the UK in the 1970s and which focused on popular cultural and sub-cultural forms and practices, have maintained an ethnographic strand over the years (Hebdige 1979; Muggleton 2000), with researchers like Paul Willis (1977; 2000) maintaining that it is the ethnographer's task to reveal the 'lived meanings' that individuals give to their lives and activities. Willis (2000) makes a distinction between qualitative data and 'quality data'. The latter, he maintains, can only be achieved over time (in the field), not just by dipping in and asking some questions. The sociologist Robert Dingwall (1997), would agree to a certain extent that the method of 'hanging out' or participant observation, in its different guises (see Denzin 1989; Fielding 1994), is preferable to interviewing because he considers that while the latter constructs data, the former 'finds it': 'observation shows us everyday life being brought into being' (Dingwall 1997: 61). However, along with sociologists such as Atkinson (1990) and Denzin and Lincoln (1994) I would argue that it is difficult to make hard and fast distinctions between ethnography and other forms of qualitative inquiry, as I have already intimated in the discussion of performance ethnography and ethnodrama in Chapter 2 and elsewhere (Thomas 2003).

A number of factors contributed to the blurring of the borders between the disciplines of anthropology and sociology: feminist (Stanley 1990) and postmodernist (Clifford and Marcus 1986) challenges to traditional ethnographic accounts on the one hand; the rise in status of qualitative methods in British sociology (Hammersley 1992) on the other, coupled with sociology's subsequent turn to culture in the 1980s (Brewer 1994). The blurring of ethnographic research has increased further with the expansion of participant observation studies into other research domains where qualitative research had previously been frowned upon as being lacking in objectivity, such as political science, law, medicine and psychology. The boundaries between qualitative research and ethnography appear to be decidedly fuzzy round the edges. Moreover, just as I suggested earlier that it is perhaps no longer viable to talk about 'the body', so the diversity of ethnographic studies in a range of disciplines suggests that we need to think in terms of 'ethnographies', rather than the singular ethnography (Stanley 1990; Thomas 2003; Thomas and Ahmed 2004).

Having said this, it should be noted that across and within the traditions of anthropology and sociology there have been debates and disagreements about what the method of ethnography entails and what its end product should look like. As Van Maanen (1988) points out, anthropologists have mostly studied cultures other than their own, which are small-scale, remote and rural, spending long periods of time with the community and often revisiting the area from time-to-time in later years. Sociologists, on the other hand, have generally investigated aspects of their own urban culture, which in some small part is known to them and mostly without them having to learn to communicate in a different language. Anthropologists were immediately recognizable as 'strangers', standing out like sore thumbs in the unfamiliar cultures they were studying. In his study of Balinese culture, for example, Geertz (1975) mentions how he perceived of his own body as being large and clumsy in contrast to the quintessentially, kinaesthetically aware Balinese (see Bateson and Mead 1942 on this), who constantly stared past him as if he were not there when they encountered him.

However, as suggested above, these stereotypical distinctions do not hold much water across the board today and to a certain extent they did not necessarily do so historically either. As Daniel Miller (1995) has shown, there has always been an interest within anthropology, albeit a minor one, to explore the reflexive responses to 'modern' life as opposed to uncovering the life of so-called 'exotic' pre-modern cultures. The interest in describing and understanding the conditions and processes of modern life has become paramount within contemporary anthropology. Miller argues that this should not simply be seen as a change in anthropology itself, because that theme was always present. It also needs to be seen in relation to a shift in 'the consciousness of the peoples we study' (1), who position themselves in relation to modernity.

At the same time, sociologists, just like anthropologists, do not necessarily merge into the background and can also be an object of curiosity in the eyes of the members of the social group they are investigating. This is precisely what happened to me and Lesley Cooper when we were engaged in an ethnographic study which explored the (many) social dancing activities of people over 60 years of age in south east London and Essex in 2000–2001 (Thomas and Cooper 2003). For example, we had to convince participants in a regular community event that caters for older Irish people from across London that we were not the 'social workers' they assumed us to be before they would even think of talking to us. Although the participant observers effectively spoke the same language as the participants and to all intents and purposes dressed in a similar way, we nevertheless stood out as foreign bodies in this particular social setting, as was evident by the questioning looks we received as we walked into the hall. In effect, and somewhat like Geertz, we were treated as external (anthropological) others by the participants. It did not matter that we were engaging with aspects of our own cultural milieu or that our 'gatekeepers' who were trusted by the participants had tried to clear the ground for our entry into the community. The observers, in effect, became the observed (Stocking 1983). Geertz found that he was treated more informally by the local Balinese villagers after he and his wife ran for their lives along with them when the Javanese police raided an illegal cockfight they had been watching. Geertz's complicity with the local villagers as opposed to the authorities provided an opening for the basis of his 'rapport' with the locals and an entry into the culture. In a similar but less dramatic vein, Lesley and I became more acceptable to the participants taking part in the Irish community event when we took to the dance floor to join in the set dances with considerable gusto, but without a great deal of skill, much to the amusement of the assembled audience and the dancers. In this instance, and in other aspects of the study (Thomas 2004), our embodied participation opened up the possibility of rapport.

Rapport is often assumed to be the 'ideal condition of fieldwork'; having gained access, and effectively moving from the 'outsider' to the 'insider' position, the ethnographer can then get on and do the research from the inside (Marcus 1998: 106). However, this assumption has been called into question (not least by Geertz himself) with the contemporary challenges to the ethnographic enterprise. Access is not restricted to the moment of gaining entry into the ethnographic site under consideration, rather, as Hammersley and Atkinson (1995) point out, it is a continuous process of negotiation.

What is important in both cases indicated above is not the assumed entry point into the culture or group; rather what they capture is the necessarily embodied character of both the field and the researcher in ethnography (Coffey 1999). Moreover, there are times when the ethnographer does not want to gain rapport or indeed to be noticed, which again may involve some redefining of the researcher's everyday bodily experience. For example, in her study of teenage girls in south London in the late 1980s, the Swedish ethnographer, Helena Wulff (1988), found that she had to dress in jeans and sneakers and walk like a man to be safe on the streets and not attract attention to herself as a woman so that the cars would not stop and try to pick her up. Wulff also recalls how her landlady in south London taught her 'that if there was a strange character approaching, you had to turn without appearing to avoid him' (see also Okely 2007: 68).

Ethnography as a method has remained the backbone of social and cultural anthropological research, since Bronislaw Malinowski and Franz Boas revolutionized the discipline in the early part of the twentieth century by inviting student anthropologists to get out of their armchairs and into the field of inquiry to 'figure out' as Geertz has put it, 'what the native says and does' (1974: 30). Boas' and Malinowski's experiential approach was in sharp contrast to earlier speculative theoretical approaches to the social history of a culture based on second hand accounts or the 'count and classify' model based on formal interviewing (Van Maanen 1988: 16–17). The power of the experiential model of ethnography, which is credited to Malinowski and Boas, has stood the test of time within anthropology in particular, although their modes of analysis and findings have been subject to scrutiny and criticism in the intervening years (Stocking 1983). The publication of an author's first ethnographic monograph, until recently, for example, has stood as essential proof of the author's rite of passage from the status of the fledgling apprentice anthropologist to that of the professional (academic) practitioner.

SOCIOLOGICAL NATURALISM AND REALISM: ENTERING THE FIELD

Sociology, too, has a kind of textbook history of ethnographic fieldwork. The seeds of its gestation are usually located in terms of the late-nineteenth century social reform movement in Britain with reference to the studies of Beatrice and Sydney Webb, Henry Mayhew and Charles Booth, who used observation as a mode of investigation in addition to social survey methods to record the squalid life and work conditions of the urban poor. W. E. B. Dubois in the US in the 1890s, also conducted fieldwork *in situ* and interviewed people in the slums of Philadelphia using a structured questionnaire. Such studies, however, were the exception rather than the rule as Van Maanen (1988: 17) points out, as 'most fieldworkers lingered in the sweatshops and lodging houses of the day only as long as it took to ask a hasty question or two'.

The main driving force behind the development of sociological ethnography is usually attributed to the Chicago school (Atkinson 1990; Bennett 2000). Robert Park, W. I. Thomas, Ernest Burgess and others at the University of Chicago in the late 1920s and the 1930s invited their students and assistants to turn their sociological gaze towards the underside of urban life, the dance halls, the hobos and street corner gang cultures, by examining them as if they were 'anthropologically strange' (Garfinkel 1984), despite the fact that the groups under study did in fact belong to the same large-scale society as the students and sociologists.

Thus, early studies employed first-hand accounts of urban settings and in effect, utilized the same methods as anthropologists were using to study the 'exotic' American Indian cultures of the day. Park, for example, as a former journalist, preferred a documentary reporting method. He considered that by digging deep into the social milieu in question, it was possible to get to the 'real story' and that, in a positivist vein, the resulting facts would speak for themselves.

As well as studying deviant subcultures, involving data gathering *in situ*, a second research strand of this school involved community studies, whereby teams of researchers, under the guidance of a lead researcher, would go into a town or a community and gather as many facts as possible in a short space of time. This kind of research did not involve close participant observation over an extended period by the team, although the lead researcher might live in the community for a while. In this strand 'hanging out' in the community was generally superseded by the quest for data gathering and reporting findings. The scope of the Chicago school was broadened in the 1940s by the work of Everett C. Hughes who encouraged his students to study more privileged groups such as doctors, teachers and policemen. The classic example of this is perhaps the study of how student medics become doctors, *Boys in White* (Becker *et al.* 1961). Nevertheless, the study of the underdog by and large remained a dominant focus of ethnographic study in sociology.

Although ethnography has diversified considerably over the years, as suggested before, the legacy of the Chicago school looms large in the sociological ethnographic imaginary. The ethnographic studies that emerged from the Centre for Cultural Studies at Birmingham in the 1970s, for example, drew heavily on the Chicago school's focus on the underdog (Denzin 1989) and other ethnographies, such as Bennett's (2000) study of music cultures in north east England, also note the influence of that tradition.

The ethnographic reporting employed in the 'historic phase' of the Chicago school (Atkinson 1990) has clear parallels with the conventions of the literary genre of naturalism, particularly through the legacy of Park and his students. However, in his book on the history of the Chicago school, Martin Bulmer (1984) argues against assuming an all too easy fit between naturalism, journalism and the early Chicago ethnographies. This is because the latter were applying more systematic research procedures to the study of the underside of a complex, mobile, modern society: '[The] Chicago sociologists conceived of themselves as scientists' (1984: 96–7).

However, discussions of the methodological aspects and problems of conducting field research, as Bulmer notes elsewhere (1983), do not pervade the early classic texts inspired by Park, such as *The Hobo* (Anderson 1923) or *The Gang* (Thrasher 1927). The publication in 1983 of an early unpublished paper by Paul G. Cressey based on research conducted for his classic Chicago study, *The Taxi-Dance Hall* (1968), which was first published in 1932, sheds light on 'the development of field research methods at Chicago in the 1920s' (Bulmer 1983: 96) and is pertinent to the discussion on the researcher's bodily positioning in the research process.

The taxi-dance halls were so called because the male patrons who frequented these 'closed' establishments, who were primarily itinerant ethnic minority immigrant workers, bought tickets to dance with young girls who were paid to dance with them. *The Taxi-Dance Hall* study was based on observational data which Cressey and four other assistants gathered between 1925 and 1929. As such, it stands as an early example of 'observational team research' (Bulmer 1983: 97). Cressey does not dwell on the methodological procedures in the monograph, although he does inform the reader that formal interviews did not yield much data in this setting and that the interviews with the proprietors of the closed dance hall were unhelpful. However, in the 1927 paper, which was subsequently published in 1983, Cressey sets out the methods that he used for the taxi-dance hall study, including examining the role of the social researcher. Here, he analyses and elaborates on Georg Simmel's (1950 [1908]) classic essay on the role of 'stranger' in sociology. For Simmel, the social researcher is a part of the community or group under investigation and at the same time, s/he remains an outsider. The 'near-far' relation facilitates the possibility of generating confessional rapport on the one hand and objectivity (through distance) on the other. Here, again, we are confronted with an oscillating shift in vision (near/far) and bodies (researcher/researched). As the chapter develops, it will become clear that this remains an issue or a problem in the sociological imaginary.

In his paper, Cressey (1983) discusses three different types of social relationships which are used by social researchers in conducting 'case studies', which involve the physical and social relation of the researcher to the research field in different ways. The first is the use of 'intimates' (friends, family) to record information, which he notes is rarely used. The spatial relations between the researcher and the researched are close, constituting extreme familiarity with your own sort, so to speak. The second social relationship, following Simmel, is the 'sociological stranger'. Here, the researcher is viewed as 'a product of mobility in that he is physically present but yet culturally distant from the group – and yet a part of it' (Cressey 1983: 104). The physical proximity of the researcher to the group exists but this does not mean that s/he is culturally part of it. Thus, the near/far relation of the researcher to the research field has been recognized as a problematic feature of ethnographic research over the years and not just since the 1960s, as is often assumed. The third typology, the 'anonymous stranger', refers to the use of anonymity in social research, which had received no 'formal consideration . . . for the purposes of research' at the time of writing, according to Cressey (109). The role of the anonymous stranger refers to 'the relationship of casual acquaintances who meet anonymously, and under the cloak of anonymity exchange mutual experiences and sympathies' (104). These anonymous relationships, Cressey notes, are a common feature of 'our mobile city life' (ibid.). The transient, unmarked physical presence of the researcher in this relationship offers the possibility of generating chance acquaintances with other transient outsiders to glean information which would not be elicited otherwise because of the social situation. The sense of bodies in relation to space and place is palpable in this image of strangers seemingly happening on each other in the somewhat (morally) dubious context of the taxi-dance hall, although as readers we already know that at least one of the strangers is purposefully inducing these casual acquaintances with sociological intent.

Cressey draws on his covert observational fieldwork of the closed taxi-dance hall to assess the methodological strengths and pitfalls of the role of the anonymous stranger. In contrast to his teacher Burgess, who encouraged and developed the life history method, which was a cornerstone of the Chicago deviance studies, Cressey's paper offers a very early account of the role of participant observation in field research, which much later would come to be known as ethnographic methods. As opposed to adopting a social-psychological viewpoint, Cressey considers the processes of 'typical social situations and rules and procedures' in settings like the closed taxi-dance hall where 'lonely unattached men come to pay a young girl to dance with them' (110). In this sense, perhaps, he is a forerunner of Goffman's approach to interaction studies. Cressey is also aware that whilst adopting the role of the anonymous stranger is useful in the transient world of the taxi-dance hall, it would not necessarily work in other social settings. Hence, he implies that the social context may provide the key to the most effective approach to be adopted by the social researcher.

In contrast to some of his contemporaries, Cressey advocates a non-judgemental stance towards the people who inhabit the world of the taxi-dance hall. The moralizing tones of the authors pervade the early Chicago studies of dance halls and as such, exemplify 'the moral continuum that characterizes the Chicago school' (Dubin 1983: 91). Cressey, however, was seeking to examine the social situation in terms of 'accurately describing and analyzing the processes operating within' (86) without imputing his own values on the social activities and customs of the clients and dancers in the taxi-dance hall setting. In order for people to be willing to tell their stories, a certain rapport has to be established between the researcher and the individual. It is the researcher's task, Cressey argues, to create the right situation for this to happen.

Thus, Cressey's paper on the role of the stranger in sociological research provides fresh insights into the development of field research methods in the Chicago school of the 1920s (Dubin 1983) and in certain respects questions the all too easy analogy between the Chicago sociology and literary naturalism.

Atkinson (1990) maintains that his interest in drawing parallels between the early Chicago school sociologies and naturalism in literature is largely 'a heuristic one' as they share a number of stylistic features. The perceived traffic between ethnography and other literary forms is not restricted to the Chicago school. In his chapter on 'Blurred Genres', Geertz (1983), discusses the ideas and tropes from the humanities that found their way into the social sciences in the 1970s. Key examples are found in the use of 'game theory' and the dramaturgical analogies in the work of Goffman, or in Victor Turner's (1982) approach to ritual theatre, which were discussed in earlier chapters. This traffic, as Geertz also notes, was not simply one-way; the humanities also picked up certain kinds of attitudes or ideas from the social sciences, such as motives, authority and persuasion.

Thus, crossovers and parallels occur at different moments in the history of ethnography and also within particular timeframes. Van Maanen (1988), for example, proposes a tripartite typology of the narrative conventions which are associated with ethnographic writing about culture: 'realist', 'confessional' and 'impressionist'. These three types of ethnographic 'tales' can be found to exist in one ethnographic study or at different points in the history of ethnography. The forms are not simply conceived as evolutionary, despite the fact that Van Maanen recognizes only too clearly that ethnography has undergone many changes since the early Chicago studies. Rather, he maintains that the different styles can be found to co-exist within different points or moments in the ethnographic canon.

REALIST, CONFESSIONAL AND IMPRESSIONIST ETHNOGRAPHIC TROPES

The 'realist tale' is the most common type to be found in ethnography. Here the teller of the tale (the researcher) stands outside of the social setting looking in (or down), objectively reporting in 'a measured intellectual style' (Atkinson 1990: 33). It is thus characterized by ethnographic authority. Although the writer is invisible, the writing is drawn from experience (in the field) so that 'everyday details' permeate the text. These fine details give the text the 'I-was-there' impression, giving credence to the researcher's keen ethnographic 'eye', while absenting his/her personal ethnographic 'I'. Thus, the realism of the account 'is conveyed through textual representations of the concrete, the local, the detailed' (ibid.). This, in turn, helps to substantiate the methodological approach taken. In realist tales the participant's point of view is taken as important in that it gives voice to the individual's perceptions, experiences and understandings. In so doing, it also provides for the analytic framework for the ethnographic story (Van Maanen 1988: 66–67). However, the interpretation does not stem from the 'ethnographic natives', which are essentially ideal types, but rather from the ethnographer. In other words, the classic realist account is a second or even third hand construct based on what might be termed as a 'view from nowhere' (Bordo 1993) with the authorial voice of the disembodied observer skating effortlessly across the ethnographic landscape. However, the act of seeing or looking is most clearly an embodied activity, which, here, is couched in terms of methodological distance under the guise of objectivity.

The 'confessional tale', as Van Maanen notes, has been viewed as a reaction to realist conventions which are difficult to sustain, particularly in regard to the claims to scientific objectivity which often pervade realist texts. In effect, in the confessional mode, the author of the ethnographic tale is brought in from the cold. The view from nowhere becomes a view from somewhere. The realist tale is told through the eyes, and the unexplicated concerns and practices of the author. In confessional accounts, the first-hand experiences of the author are emblazoned in the text. The confessional account seeks to 'demystify fieldwork' by showing how the technique is practised in the field (Van Maanen 1988: 73). The author is not a passive observer, as in the realist text; rather s/he is actively engaged in learning from, understanding and interpreting the life of the culture or group under study. In so doing, the confessional ethnographic 'I' offers a challenge to the disembodied voice of the realist text. However, confessional accounts, according to Van Maanen, are not necessarily diametrically opposed to realist tales: 'They typically stand beside them, elaborating on the formal snippets of methods that decorate realist tales' (75). Moreover, in the confessional account, it is the authorial voice of the ethnographer that is being (re)presented and not the native's point of view. The dangers of 'going native' in confessional accounts are often tempered by engaging aspects of realism. The mixing of the confessional with the realist results in 'tacking back and forth between an insider's passionate perspective and an outsider's dispassionate one' (77).

The confessional tale has become an institutionalized form of ethnographic writing in that the researcher is a participant in the ethnographic setting. It aims at lifting the veil of secrecy around how knowledge is solicited and produced in the practice of fieldwork. Some confessional accounts do aim to maintain the objectivity of 'social science', as opposed to seeing this as a fiction that functions as truth (Rabinow 1986). However, increasingly, as Van Maanen (1988) points out, confessional writers are not interested in sanctioning scientific orthodoxy. This is evident in feminist and postmodern critiques of ethnographic authority and in recent debates around the rise of autoethnography, which pose stylistic and linguistic challenges to the orthodoxies of social research (Seale 1999; Ellis and Bocher 2000; Denzin 1997).

The import of the critique of ethnographic authority is that 'social facts' can no longer be seen as 'neutral, objective, observable facts' (Van Maanen 1988: 93). 'Rather, social facts, including native points of view, are human fabrications, subject to social inquiry as to their origins' (ibid.). From here it follows that 'fieldwork is an interpretive act, not an observational or a descriptive one' (ibid.) which requires reflexively examining one's own taken for granted assumptions and biases and from there, adopting an approach to understanding which involves 'a continuous dialogue between the interpreter and the interpreted' (ibid.). This entails that there is a body or bodies in the text. The embodied subjectivities of the researcher and the researched are necessarily brought into the methodological frame through the research context.

The third ethnographic typology is the 'impressionistic tale'. In this mode, accounts of some dramatic or unknown world attempt to draw the reader into the perspective of what the author saw, felt, heard and so on. The author aims to enable the audience to place themselves in the context of the fieldwork, to give them a ringside view of the events, almost being able to touch and feel what is going on. Impressionist tales capture the scene at the time in a moment, not a forever, kind of way. Like the nineteenth century impressionist painters (e.g. Monet, Pissaro, Renoir, Van Gough), ethnographic impressionists try to capture the presentness of events or activities by means of an impressionistic view of what is going on, not a realistic, holistic picture. Impressionistic ethnographies, then, deal in partial truths. Just as the nineteenth century impressionist artists sought to confront the academic tradition by using their artistic materials in innovative ways, so impressionist ethnographers aim to 'startle' their readers by their use of 'striking stories' drawn from recollections of fieldwork experiences (Van Maanen 1998: 101–2). The impressionistic tale focuses on the extraordinary events, which make an impression on the ethnographer. Royce's (1980) story of the erupting social drama between particular members of a Zapotec dance group from Juchitán in Mexico, which was discussed in Chapter 2 is a good example of an impressionist tale, as is Geertz's (1975) famous 'Deep Play' essay of the cockfighting event in Bali which was alluded to earlier in this chapter. Wacquant (2004) also has a highly impressionist tale in his study of boxing, in the third section of his ethnography, which will be discussed later in the chapter.

Van Maanen considers that impressionist tales 'present the doing of fieldwork rather than simply the doer or the done' (1988: 102). They are recollections from the field which made an impact on the author and 'the story itself, the impressionist tale, is a representational means of cracking open the fieldworker's way of knowing so that both can be jointly examined' (ibid.).

Generally speaking, impressionist tales read like a novel or short story. The literary approach is a distinguishing feature of this sub-genre from the other forms of ethnographic reportage. Van Maanen notes that impressionist tales do not sit easily alongside realist or confessional tales because of the specificity and eventful character of the former. Impressionist tales imply that the learning of a culture takes place through the extraordinary, not the everydayness of cultural events and practices, which is anathema to many researchers (Geertz 1983). Whilst the 'episodic, complex, and ambivalent realities' of impressionist tales may be too neatly packaged from a realist or confessional perspective, impressionist tales cast a shadow over the implied fixity of cultures within realist and confessional accounts, by their openness and contingent and interpretive character of social situations (Van Maanen 1988: 119).

Whilst Van Maanen recognizes that these three forms of ethnographic reporting could be viewed as developmental, it would be a mistake to treat them as such, as they exist alongside each other and there is ultimately no fixed 'temporal connection' (126). Ethnography has become more complex, as Denzin (1997) demonstrates, but older forms such as realism still exist in tandem with new experimental reflexive models, although perhaps thoroughly purged of their more problematic conventions. The very notion of culture itself, as indicated before, has undergone something of a transformation since the 1980s which also has implications for contemporary realist and confessional tales that take this on board. In the hands of the now not so 'new' (postmodern) ethnographers such as Marcus (1998), for example, culture emerges as a contested construct, which is fluid, contingent, emergent, partial, ambiguous and 'messy' for both the ethnographer and the social agents of study. These shifts in people's 'lived cultural experience' should be represented in ethnographic representations of culture. Fixing cultures in terms of history by means of always speaking of the present, along with assuming a closed and fully knowable social world, as told by a disembodied observer, do not reflect these lived experiences, if indeed they ever did (see Marcus 1998: 33–55 on methodological holism in contemporary realist ethnographies).

MORE ETHNOGRAPHIC TALES

Van Maanen recognizes that his three ideal typical ethnographic genres do not cover the totality of the ethnographic landscape. Like Denzin and Lincoln (1994), he points to the fruitful and challenging development and diversity of contemporary ethnography. He proposes 'four additional forms of ethnographic expression' (1988: 127) as possible additions to his tripartite model, in light of more recent interventions in the discourses of both ethnography and culture. 'Critical tales', as evidenced in the approaches taken by Willis (1977) and Marcus and Fischer (1986) for example, denote a concern to take account of the larger social landscape which impacts on the ethnographic setting. This form has often been grounded in a Marxist-oriented framework. The tellers of the second form, 'formal tales', are also concerned to generate a more generalist framework, 'to build, test, generalize, and otherwise exhibit theory' (Van Maanen 1988: 130). This mode directs attention to a more narrowly drawn ethnographic focus, under such 'labels' as ethnomethodology, as exemplified by Garfinkel (1984), semiotics, conversational analysis, socio-linguistics and includes various structuralist approaches; where, despite differences, the aim is to make generalizations 'through inductive and inferential logic' (Van Maanen 1988: 130). The theoretical framing of different formalist, structural approaches, such as those advanced by Lévi-Strauss (1978) or Douglas (1970) for example, often conflict with the time-honoured practice of 'doing' fieldwork, as the research field runs the danger of becoming a testing-ground for the theory. The third form, 'literary tales', borrows much from what has been labelled as the 'new' journalistic writing, as encapsulated in the work of Tom Wolfe, for example (Denzin 1997). For Van Maanen, the 'distinctive marker of the literary tale lies not so much in the fieldwork style adopted, as in the 'borrowing of fiction-writing techniques to tell the story' (1988: 132). The fourth form, 'jointly told tales' refers to the trend in postmodern ethnography for 'dialogic and polyphonic authority in fieldwork representations' (136). This mode is exemplified in Clifford's (1988) critique of ethnographic authority and Marcus' (1998: 79) pursuit of a 'multi-sited ethnography', which moves away from the conventional single, localized, situational ethnographic research model in favour of examining 'the circulation of cultural meanings, objects, and identities in diffuse time-space'.

As Van Maanen's discussion shows, 'there is no sovereign method for establishing fieldwork truths' (1988: 138). The truth of the matter is that ethnography, as with other qualitative methods, is 'messy' (Marcus 1998).

> Self-understanding is not the point of fieldwork as confessionalists sometimes suggest. Nor is the brilliant but necessarily objectified, representation of another culture the endpoint. Impressionist tales dance around both poles and inform, elucidate, amuse, and invoke in useful ways. Their open-endedness is their strength, for meaning can be worked on again and again and few readers are excluded. But even here, impressionist tales can be used in many ways, not all of them good.
>
> (Van Maanen 1988: 138–9)

The three substantial ethnographic tropes of realism, confessionalism and impressionism, as already noted, are not simply to be found in the chronology of ethnography, they can be evidenced together in a given period and may also be found in one ethnographic text, which may or may not privilege one of the other modes as its primary tale. But as has been noted earlier, whilst the voice of the researcher *may* be heard in the text, his/her embodied engagement in the field is largely implied and not explored in any depth (Okely 2007). This cannot be said of Wacquant's boxing ethnography, *Body and Soul: Notebooks of an Apprentice Boxer* (2004).

CONFESSIONALISM, REALISM AND IMPRESSIONISM IN ACTION: A TALE OF TOTAL IMMERSION IN THE SPORT OF BOXING

The first sentence in the Preface to Wacquant's ethnography, which is divided into three main sections, gives the reader a taste of things to come and signals that this may not be a typical ethnographic text. *Body and Soul* is described as, on the one hand, a sort of 'sociological-pugilistic *Bildungsroman*', which tracks the researcher's 'personal *experience* of initiation' into the 'bodily' craft of boxing and on the other hand, as 'a scientific experiment'. There is a phenomenological influence at work through the focus on 'lived-experience', which may appear to be at odds with the notion of a (sociological) scientific experiment. The *Bildungsroman* is a nineteenth century literary form which centres on the moral, psychological and intellectual development of the main protagonist or hero. It is essentially a novel of self-development and realist in form, as exemplified by, for example, Charles Dickens', *Great Expectations* or Charlotte Brontë's, *Jane Eyre*. The realist novel is an epic dramatic form, involving multiple plots, a variety of precise social settings, and strong political and social contrasts, with an enlightened or hopeful ending (Caute 1972), all of which are visible in Wacquant's book. The Marxist literary theorist, Georg Lukács, argued that great literature, as exemplified by the realist tradition, should not only reflect the total reality of the time but should also attempt to transcend it. In order to achieve this ideal form of critical realism, he recommended the *Bildungsroman* (Swingewood 1998).

In Wacquant's ethnography, realism, confessionalism of a sort, impressionism and readily acknowledged literary licence, are in abundant evidence, with the researcher's total bodily immersion in the 'culture of bruising' (Early 1994) in a South Side Chicago boxing gym playing a central, starring role. Wacquant's ethnography weaves the more formal, realist aspects of the social ecology of the ghetto through his embodied narrative journey to become a fully-fledged boxing body, beginning from his apprenticeship through to his first major fight in the famous Chicago Golden Gloves some three and a half years later, which is a glorious example of an impressionist tale, shot through with the characteristic detail of the realist tale and a measure of confessionalism. The tropes of realism and verisimilitude are evident also in Wacquant's concern to relay the words of his fellow pugilists and trainer in their own spoken idiom as it were, rather than 'translate' them into formal English. For example, DeeDee the trainer's 'implicit and collective pedagogy' of boxing, according to Wacquant, consists of giving advice in short negative remarks to the boxers which he considers does not need any explanation, such as: 'It's easier than counting one-two-three'; 'There ain't nuthin' to explain, what you want me t'explain?' 'We'll see later, *just box*' (2004: 103, my emphasis).

Boxing and particularly prize fighting is a hugely popular spectator sport which, as its fans are only too well aware, gives rise to a great deal of censure (Woodward, Kath 2008). This erupts into periodic public outbursts that call for a ban to be placed on the sport because of, for example, irreparable damage to a boxer or even the death of one, as a consequence of long term fight damage or a single deathly fight. Moreover, boxing has had a reputation of attracting people from the criminal edges of society who have engaged in dubious practices such as fight-fixing and the gross exploitations of the fighters. Despite its morally dubious reputation throughout great swathes of its history, it has continued to be popular with the public (Sammons 1989). Boxing is a highly individualistic, 'gladiatorial' physical contact sport where the aim is to hurt or injure your opponent and if possible, to have him counted out or knocked out before the end of the fight; 'the brain is the target, the knockout the goal' (Oates 1987: 93). The other related aim is to defend oneself against being hit, so as not to incur physical damage. Having noted this, it is also the case that the 'modern sport of boxing' has gone through a 'civilizing process' from its period of development in the seventeenth century to the mid-nineteenth century, where it was, by contemporary standards, 'an extremely violent, brutal, and bloody activity' (Sheard 1997: 35).

The history of boxing is deeply rooted in class, gender and racial delineations. Often referred to as the 'Manly art' (Wacquant 2004), it is a sport which is overwhelmingly practised by adolescent males and young men from impoverished socio-economic inner-city communities, which are ethnically delineated, and which change over time (Weinberg and Arond 1952: 460). In the history of boxing in Chicago, for example, the 'ethnic succession of boxers' stemmed from the South and West sides and these corresponded to the different ethnic groups that inhabited these areas at particular points in time. When one group became more prosperous and moved out, the next group of immigrants moved in. 'First, Irish, then Jewish, then Italian were most numerous among prominent boxers' (ibid.), after which, African Americans and more recently Latinos, rose to precedence.

Several new studies on boxing offer evidence of a further shift in recent years (Heiskanen 2006; Woodward, Kath 2008). More women, middle class and white-collared workers are learning the craft of boxing and can be seen to work out in the gym alongside the core participants who have traditionally been attracted to boxing (Heiskanen 2006). It is well documented that disadvantaged youths from ethnic minority groups often see the sport as an opportunity to get out of the ghetto and into the hall of fame and fortune (Sammons 1989; Wacquant 2004; Heiskanen 2006). The history of boxing greats, such as Jack Johnson, Joe Louis, Sugar Ray Robinson, and Mohammad Ali, whose pictures are lined up on the boxing gym's hall of fame wall, offer inspiration and a constant visible reminder of what is possible.

Much has been written about prize fighting and famous champion boxers over the years, by for example, sports writers like A.J. Liebling (1956), and by the boxers themselves, too. The sport has also inspired many writers such as Norman Mailer (1975), Joyce Carol Oates (1987), Thomas Hauser (1991) and Gerald Early (1994) to put pen to paper on the subject. It is evident that boxing, has also touched the popular cultural imagination in the cinema over the years as over 150 boxing films have been made since 1930 (Grindon 1996) and the male boxers play the central role. There is always an exception to the rule and *Million Dollar Baby* (2004) has Hilary Swank playing the leading role of an ambitious female boxer who is concerned to be taken seriously as a fighter, which perhaps reflects the recent incursion of women into the sport, although her character ultimately dies for it in classic boxing film style.

The boxing film often charts the rise and fall of the boxing protagonist, explores his coping mechanisms in dealing with his inevitable diminishing bodily skills and his relations inside and outside of the magic circle of 'the ring'. The main title of Wacquant's monograph, *Body and Soul*, is interesting in this connection, as it obliquely refers back to two major boxing films of that title made in 1947 and 1981, with which the author would have been familiar. These two films, according to Leger Grindon (1996: 54), acknowledge the fundamental issue that haunts the boxing film per se, 'the conflict between body and soul established by a tension between physical reality and psychic experience, the material and the spiritual'.

Whilst the western cultural tradition of thought mostly privileges consciousness over the body in debates on divisions between mind/body, 'the boxing film foregrounds the body and particularly the male body' (ibid.). Wacquant (2004: 17) considers that 'the boxer is a *live gearing* of the body and the mind that erases the boundary between reason and passion'; the boxer *is* his body. As such, the boxer may be seen to exemplify the fact that, following Marx, 'the social agent is before anything else a being of flesh, nerves and senses . . . who partakes of the universe that makes him, and that he in turn contributes to making, with every fiber of his body and his heart' (vii).

This 'carnal dimension of existence' is particularly evident in boxing, according to Wacquant, but it is also part and parcel of our everyday life. For Wacquant, the challenge of a carnal sociology, which is so often lost or silenced in established social science approaches, is to encapsulate 'the taste and the ache of the action, the sound and the fury of the social world' (ibid.). The way to achieve this is through 'an initiatory immersion' in the social milieu under investigation which places the investigator at the heart of the action (rather like the *Bildungsroman*). Wacquant argues that if Bourdieu's contention that we learn 'by body' and that the 'social order inscribes itself in bodies through its permanent confrontation' is correct, then, 'it is imperative that the sociologist submit himself to the fire of action *in situ*' (cited in viii).

Wacquant initially signed up at the Woodlawn Boys Club on 63rd Street to try and get some first-hand, on the ground, information and understanding of everyday life in the black American ghetto, which would have otherwise been unavailable to him. This gave him a legitimate reason to 'hang around' the gym and meet young men from the local area. 'Hanging around' the gym soon turned into a more intensive 'hanging out' *with* the boys as he surrendered himself to the rigours of learning the 'Sweet science' of boxing, to echo Liebling's term (1956). In turn, this contributed to this white Frenchman being viewed as one of the guys (*'brother Louie'*) in the all black gym and perhaps more importantly, as a member of the professional trainer's core group, 'one of DeeDee's boys' (11), both within the club and with other trainers and athletes outside of it. Wacquant also notes, as others have done too (Heiskanen 2006; Woodward, Kath 2008), that the gym is a democratic level playing field in many ways if one is seen to buckle down and submit to the discipline of learning the craft. As will be discussed later, this is somewhat easier for men than women. In the potentially dangerous and impoverished locales where the boxing gyms are mostly located, being known for being able to take care of yourself through boxing, can in certain instances lower the risk of being attacked and provide for a safer passage in the ghetto (Sugden 1996; Woodward, Kath 2008). Indeed, some boxers cite the fact that they wanted to defend themselves as a major reason for joining the gym in the first place, and praise its role in helping to keep them out of trouble or getting into bad company in the street (Weinberg and Arond 1952; Heiskanen 2006). Wacquant's (2004) insider status also facilitated a degree of safety on the streets of the ghetto, outside of the sanctity of the gym. Just to make sure, he was given a present of a can of mace to carry with him and give to his wife in order to defend against the very real possibility of a street attack.

Wacquant trained for three to six sessions a week for three years with amateurs and professionals in the Woodlawn gym. He also attended other gyms and some 'thirty tournaments and boxing "cards" ' ... in his capacity as 'gym mate and fan, sparring partner and confidant, "cornerman" and photographer' (2004: 4). All of which 'earned' him access to the front and back stages of 'the theater of bruising'. Wacquant's total involvement was such that he even thought of turning professional so that he could just stay there for as long as possible in the boxing fraternity and give up on his academic career. He describes his ethnographic immersion approach as one of 'observant participation' which builds on and develops the idea of participant observation. Observant participation places emphasis on being observant through 'doing'. Wacquant mentions in a field note that 'PB' (Bourdieu) had expressed some concern that he was 'being seduced' by his object of study. Wacquant goes on to note: '*if he only knew*: I'm already way beyond seduction' (4). Hence, his acknowledgement of the fact that he had indeed gone 'native'; a subjectivist move which is viewed as dangerous to the validity of social research from an objectivist standpoint, because of the intense personal involvement in and subjugation to being *in* there.

Bourdieu (2003) is highly critical of what he considers is 'the pseudo-radical denunciation of ethnographic writing as "poetics and politics" which is characterized by postmodern ethnography and the phenomenological and ethnomethodological imperatives of dwelling on the lived experiences of the researcher'. He questions how the participant observer can be both subject and object of research at the same time; that is, 'the one who acts and the one who, as it were, watches himself acting' (281). Bourdieu contends that it is not necessary to choose between participant observation on the one hand and the 'objectivism of a "gaze from afar" ' on the other, in which the researcher remains remote from the that which s/he is studying. Instead he seeks to promote what he terms 'participant objectivation' which requires that the analysing subject, the researcher, be objectivized. 'Participant objectivation undertakes to explore not the "lived experience" of the knowing subject but the social conditions of possibility – and therefore the effects and limits – of that experience and, more precisely, of the act of objectivation itself' (282).

Wacquant considers that his lengthy immersion in the boxing gym, participating in its activities and exchanges on a routine, day-to-day basis, provided him with a unique vantage point from which to rethink his approach to understanding 'what a ghetto is generally' and the 'structure and functioning of Chicago's black ghetto . . . at the end of the twentieth century' (x). It will be recalled that this was the original aim of his research project. However, this was initially overtaken by the first challenge he set himself: to submit to being inculcated into the 'Manly art'; to learn the sport to the point where he would be accepted and have a place in the 'fraternal and competitive' social universe of the gym. This proved to be a (physically) painful, brutal and exhilarating process of engagement. Having succeeded in locating himself 'durably in the milieu', Wacquant then returned his original challenge; asking whether he could now 'grasp and explain' the black ghetto through his immersion in the specific social setting of the gym. He considers that he was able to challenge and reconstruct his approach to the ghetto through the prism of his observant participation and to reject approaches based on what he called the ' "Orientalizing" vision of the ghetto' (xi) and its inhabitants. The third challenge, relates to his learning the science of bruising: how would he be able to account ethnographically for 'a practice that is so intensely corporeal' (ibid.), and one that he was fully engaged in a physical sense? That is, how could he 'retranslate' the comprehension that emerged 'by body' into a 'sociological language and find expressive forms suitable to communicating it' (xiii), without reducing the specific characteristics that makes up the Sweet science.

This, of course, is a key concern for other ethnographers and scholars, for example, whose object of study is dance (Thomas 1995; Sklar 2000, 2001). Like boxing, dance, at least in the west, is largely concerned with expressive, aesthetic embodied practices and most dance ethnographers and scholars are dual trained in specific dance practices and in the relevant philosophical, social or cultural frameworks within which they work. Dance is viewed as a form of embodied cultural knowledge and the general ethos is that you should gain physical competency in the bodily knowledge of the dance form you are intending to analyse. Like boxing, dance is usually learnt implicitly by doing, imitation, correction and repetition until the gap between thinking and doing is closed. Wacquant's account offers a detailed discussion of the bodily habituation of the boxer until he did not think about it. Indeed, his trainer instructed him not think about it or read about it, as it had to come from the 'heart', another term which is used in boxing literature to denote good boxers. I also recall of being told in my own dance training not to think about what I was doing but just to '*move*'! Whilst boxing may not be considered 'aesthetic' as such, in the same way that dance is perhaps, it is clear that descriptions of and distinctions between, the image of a 'yokel' and the 'trickster' in prize fighting (Early 1994: 10) are about boxing aesthetics. The yokel is a fighter who is a 'puncher', who is 'expected to take punishment, then land the ultimate blow' (12) and whose style Leibling described as 'gauche and inaccurate' (cited in Early ibid.). The trickster on the other hand, as the term suggests, is expected to win by 'guile and ability' (ibid.). While the yokel is usually associated with white fighters, the latter, as Early notes, is mostly associated with black boxers, such as Mohammad Ali, Sugar Ray Robinson and Sugar Ray Leonard. It is the 'tricksters of style' who are the most highly regarded by the experts in the field. Moreover, a boxer is accorded the title of 'Sugar', according to Early, because of his sheer style and no white boxer has achieved this accolade. Joe Louis, who perhaps has been written about more than any other boxer apart from Ali and who was known as a hitter, was primarily a stylist; with Liebling commenting that Louis was 'as elegant as the finest ballet dancer' (cited in ibid.). Although there is a codified knowledge of different dance techniques, some of which are written down, such as ballet vocabulary, it is generally taught and learnt 'by body' until it becomes habitual or in phenomenological terms, 'it goes without thinking'. It is often said that ballet dancers, for example, walk (like ducks) with an extended turnout in everyday life, which is an essential feature of the form, just as Wacquant comments that he took to moving like a boxer, without even noticing it; his inculcated pugilistic bodily 'habitus', which from Bourdieu's (1993) perspective is a 'structuring structure'.

The realist character of Wacquant's text is a contemporary one, where realism and confessionalism sit side-by-side, in the way that van Maanen (1988) has pointed out. In Wacquant's monograph, however, the sociological 'I' is deeply embedded through his lived experiences; the trials, tribulations, frustrations and triumphs of learning the craft of boxing. In this way, the 'I' takes precedence over the sociological 'eye' although the text does try to oscillate between the two. As indicated earlier, the realism of Wacquant's text is revealed in the detail of his discussion of the ecology of the ghetto in the first section in particular. In this sense, he is continuing in the Chicago tradition of human ecology, forged by Park, which focuses on excavating and understanding the relations between the social structure, the (ghetto) environment and the social groups and individuals that inhabit and move through it. However, his starting point is not as an outside observer, but through his bodily immersion in the gym which is situated in the ghetto environment. It is from this vantage point that he claims he was able to both understand and critique more conventional, contemporary analyses of the social structure and complexities of living in the ghetto.

The question that arises, here, is how important is it for a researcher to totally immerse himself in the activity of boxing in order to understand the ghetto and what problems may arise from such an intense engagement? This can be made into a wider discussion to ask how important is it for the sociologist who studies dance, sport or any other activity that requires a strong practical sense of '*direct embodiment*' (Wacquant 2004: 60) to engage in that particular practice and to what degree of proficiency? I have already indicated that dance ethnographers tend to be dual trained and that embodied participation in the dance practices under investigation is generally viewed as crucial to developing new insights on the relationship between dance and culture. The implication, here, is that the embodied practice engagement, as a methodological tool, usually associated with the influence of phenomenological thinking, has contributed to a paradigmatic shift in the field. However, as

Sally Ann Ness (2004) points out, it is not necessarily the case that dance participant studies in contrast to observational studies, have led to advances in knowledge and understanding. Turning back to the Sweet science, Oates' (1987) stunning book on boxing from the point of view of a lifelong audience member demonstrates that you do not have to be a boxer in order to gain insights into that world or its practice, as the reader can almost feel, hear and smell the action from that perspective, as well as grasp the world in which it is situated. Kath Woodward's (2008: 545) insightful ethnographic study of boxing also asks: 'How far does the researcher have to be part of the regime of the body practices in order to understand them?' Woodward suggests that *not* being part of the same 'system of practical boxing beliefs' (Wacquant 1995: 88), such as the widely held unquestioned belief by boxers that the sport is 'in the blood', 'it's in you', or 'it's natural' (Wacquant 2004), can productively disrupt certain taken for granted associations of the sport, leading to new insights. While Woodward recognizes that bodily engagement in the sport provides an important point of access and 'is one of ethnography's major strengths' (551), she also argues that insider collusion can be problematic in glossing over gendered associations and differences, which in turn has implications for the analysis and the validity of the research. For example, she suggests that although Wacquant invokes the masculine culture of boxing, he ultimately assumes it, rather than explores it; after all, as he proudly notes he is 'one of DeeDee's boys'. In so doing, he (perhaps unwittingly) posits 'a universal embodiment which fails to accommodate the specificities of gender' (2008: 545). Woodward argues that 'the associations of bodily practices of boxing do have consequences for the researcher who cannot, at least in some sense, be un-situated in relation to gender classification' (546). While male researchers have to demonstrate their qualifications within the masculinities that are played out in the context of the gym, women, as Woodward notes, 'are positioned as gendered by their very presence' (ibid.) rendering them necessarily more of an outsider status, as sex objects parading around the ring for example, as a distraction to the boxer in the gym or prior to a fight, or a sort of mother figure, like the mums who regularly bring their children to the gym. In turn, this may give rise to difficulties of access, and of their having to develop strategies of engagement, which do not involve their participation. Although some women are taking up boxing as a sport, their participation is generally seen as a kind of 'soft' boxing and therefore, they are definitely not 'one of the boys'. Woodward notes that researchers are situated within 'gendered spaces' in which they both situate themselves and are constructed by 'the embodied framework of gender' (ibid.) and the research needs to take account of this. Insider/outsider research status, as suggested earlier, is not necessarily an either/or situation that is fixed for all time in the research process and has to be negotiated on a routine basis. Woodward's article thus points to the necessarily embodied character of research (in her case a gendered one), whether or not there is active embodied participation in the sport of boxing or dancing, which was a major feature of this chapter. Although she too notes that ethnographic immersion provides for an important point of access, the necessity of shifting between the near/far relation should be kept in mind (see also Ness 2004).

CONCLUSION

This chapter has explored a range of ethnographic devices, histories and practices in which the researcher's bodily presence, the participants and the embodied field are intertwined, although they have often gone unnoticed or reported. This led to the central discussion of the near/far relations between the researcher and the research, touching on key issues such as rapport and the situated role of the researcher's physical presence in the research context. The ethnographic tropes of naturalism, realism and confessionalism which Van Maanan identified as existing across time, were used to get away from the 'evolutional' or historical developmental approaches which have dogged discussions of ethnography, although it must also be said that these too are not without merit. Finally, these tropes were explored further with reference to Wacquant's study of his lived experiences of becoming an apprentice boxer in a gym which was situated in a Chicago ghetto. In so doing, questions were raised about the merits and problems of the researcher's total bodily immersion in the sport, which once again brought the chapter back to the complex near/far relations between the researcher, the researched and the field (ethnographic and sociological).

5

OLDER BODIES: PERFORMING AGE, AGEING AND INVISIBILITY

INTRODUCTION

This chapter, as the title indicates, considers the various ways that older bodies have been constructed and perceived. However, the subtitle appears to be contradictory; the notion of performance suggests that age is enacted, negotiated and presented (to others), while invisibility speaks to the idea of the unseen or unnoticed aged body from the perspective of the outside 'public' world. In this chapter, I will discuss a range of approaches to the study of age and the ageing body in academic discourse on ageing and contemporary culture, through research areas such as social gerontology, sociology of ageing and the body, and to some extent, cultural studies. However, it is important to note that social gerontology, in particular, needs to be situated in relation to bio-gerontology, which has had much influence on the field, in both a negative and positive sense. By examining these different aspects in the context of the exponential rise of interest in the study of the body, hopefully, it will become apparent that age and ageing in contemporary, late modern society may be considered as an 'accomplishment' or a 'performance' (Laz 2003) which is situated within what Arthur Frank (1996: 53) terms 'the three part relation between the society, the body and the self'.

In everyday life, we tend to consider age in terms of chronology from the date of birth as essentially a 'natural' fact. As Cheryl Laz (1998: 85) notes, '[w]hen asked "how old are you?" we usually offer the number of years since our birth'. As people go through life they are presumed to grow into and out of different phases with particular patterns and behaviours, which are commonly affiliated to particular age stages; childhood, adulthood and old age. In the latter part of the twentieth century, Peter Laslett's influential and much criticized book, *A Fresh Map of Life* (1989), argued for the need to reconsider the three stages of the life course into a new four stage model, as a consequence of demographic shifts and societal changes around retirement. Laslett's first and second stages roughly correspond to childhood and adulthood. The third age is a subdivision of what was previously viewed as old age and is now conceived as 'a period of personal fulfilment' prior to entering into the fourth age which is characterized by 'dependence, decrepitude and death' (4).

The signs of ageing are usually associated with identifiable physical changes such as, changes in skin tone, wrinkles, hair colour, and the lessening of bodily strength and control, which are closely linked to ideas of growing old. For example, commenting that an older woman looks good for her age generally implies that she appears younger physically than would be expected for her age. In a culture dominated by youthfulness, this is intended as a complement, and most recipients would be pleased about this. On the other hand, telling someone to 'act' their age is a form of mild chastisement, which suggests that the person in question is knowingly behaving in an inappropriate manner for their age. These linguistic examples reveal the taken for granted assumption regarding what age *is* for the speaker and the receiver of the communication. Age in this sense is treated as a *doing* word, which is routinely performed, negotiated and accomplished in interaction with others, and which differentiate people by socially produced age categories, which change over time. For Laz (2003), age is something we *do* and we cannot stop doing it at will. Taking the lead from constructionist theories of gender, race and class, Laz proposes that age should also be viewed as a social construction as opposed to natural or a biological fact. However, she insists on bracketing out certain thorny implications of the meaning of performance from her usage of the word in this context. She insists that there is no hidden or real self lurking underneath the performer waiting to get out, nor is the performance of age always consciously constructed and enacted. Viewing age in terms of an accomplishment or performance (as something that is done) directs attention away from a concern with individual attributions and towards one that 'involves, routine, sometimes impressive but always ongoing, recurring and collective work' (506). As such, it is important to examine the social contexts and institutional settings in which people routinely act their age. But Laz does not propose a radical constructionist framework, which contends that the accomplishment of age is shaped by social forces to the exclusion of other factors. She argues that it is also 'constituted in interaction' through which it gains its meaning and in the context of 'larger social forces' (ibid.). From this perspective, we not only perform our own age all the time, we also assign meaning to other ages 'and to age in general in our actions and interactions' (ibid.).

The phenomenology of the body or more precisely, embodiment, is central to Laz's approach to age, which seeks to overcome the binary oppositions between essentialist or foundational perspectives on the one hand and constructionist or anti-foundational perspectives on the other. Embodiment, like age, is perceived as an accomplishment; they are inextricably linked and 'mutually influential'. The way in which we perform age has consequences for our 'corporeal experience'. She sees that bodies 'are made and remade, used, altered, and disciplined and as such, are objects of practice' (507). We do not make our bodies as we so wish; as well as acting on them we respond and react to them, as they do not always do what we want them to do. As many professional dancers can testify, although they often ignore it (see Tarr and Thomas 2011), the body can be utterly recalcitrant to the wishes of the performer at times. It seems to have a life of its own which means that necessary accommodations to achieving desired bodily feats have to be made (see also Foster 1998). Also, as noted in Chapter 1, people with physical disabilities or bodily impairments, are seldom oblivious to their body and routinely have to find ways of managing and negotiating their body in their daily life. As Frank (1995: 27) suggests perhaps, '[t]he body is not mute, but it is inarticulate; it does not use speech, yet it begets it'. Frank suggests that through stories of illness, it is possible to get underneath descriptions of the body as a thing separate from the individual story teller who is talking *about* the body and to hear 'the body creating the person' (ibid.) in the story. For example, Anne Oakley's book, *Fracture* (2007), charts the story of her changing awareness and understanding of her embodied self as a consequence of her arm being broken in an accident and its aftermath, which left her unable to use her right hand. This personal engagement with and reflections on, the changes to her body and its limitations had a powerful effect on her everyday life and Oakley interleaves the personal body/self narrative with her social scientific and feminist scholarly insights into the story. In going beyond an auto-ethnographic approach, her story speaks of embodiment in general and among other things, how the self can be confronted by ageing in a dramatic way. To return to Laz's (2003) discussion of age and embodiment, it is clear that she does not go as far as Raewyn Connell's (1995) suggestion that bodies actually have 'agency', although she concurs with the spirit of Connell's notion of 'body-reflexive practice'. The construct of body-reflexive practice is based on the idea that 'interpretations of and responses to the body are situated in social relations, in interactions and in the context of institutions that serve to construct and reconstruct those relations and institutions' (Laz 1998: 508). Thus, Laz's sociological approach to age brings us back to the relation between the body, society and the interactional self.

The chapter will introduce the main approaches to ageing and the major shifting paradigms and concerns that have dominated over the years and the problems they give rise to. In so doing, it will consider the 'turn' to the ageing body in studies on ageing in the context of the factors that contributed to this shift. It will also draw on performance case studies on social and professional dancers.

DISCOURSES ON AGEING AND THE BODY

It is only possible to offer a snapshot of social gerontology, which is defined as 'the study of the lives of older people' (Gubrium and Holstein 2000: 1) and sociological and cultural studies' approaches to ageing. As with other areas already noted, there has been a surprising disregard in social gerontology to consider the impact of the ageing body on people as they get older and the meanings associated with ageing in everyday life within contemporary western societies (Faircloth 2003). This is despite the fact that this field of study has grown significantly in the US since the end of the Second World War. Social gerontology in the UK came of age in what has been called the 'golden age of the welfare state' in the period from the late 1940s up to the late 1970s (see Higgs *et al.* 2009). Jaber Gubrium and James Holstein, writing in 2000, noted that the physiological changes associated with ageing, how people deal with the consequences of these in their everyday lives and the cultural meanings that they invoke, have not featured highly in social gerontology over the years. Rather, they argue, social gerontology's main concerns have centred on issues around 'retirement' and the problems of old age, which required policy oriented responses to alleviate them, as discussed below. Alternatively, biological gerontology, for the most part, has focused on the medical, 'pathological' aspects of the ageing body. In so doing, it has tended to overlook or disregard the potential impact of the social, cultural and environmental experiences on the biologically ageing body (Vincent 2003).

The sociology of ageing, like social gerontology, as Christopher Faircloth (2003) notes, has ignored the ageing body until very recently, particularly in the US. However, for Bryan Turner (1995), this is not exactly a surprise, as he considers that the neglect of the body in classical sociology, as discussed in Chapter 1, prevented the development of a systematic sociology of ageing. He argues that sociologists have generally not attempted to comprehend the links between varying modes of embodiment, the physical processes of ageing and socio-cultural explanations of ageing. Turner considers that the sociology of ageing should incorporate the phenomenological experiences of ageing into the frame of reference to develop an understanding of what it is like to be old. Featherstone and Hepworth's extensive writing on age and ageing (e.g. Featherstone and Hepworth 1984; Featherstone and Wernick 1995; Hepworth 1995, 2003; Featherstone 2010), which draws on postmodern and deconstructionist theories, has had a considerable impact on the development of interest in socio-cultural approaches to ageing bodies. Their articles and books in the 1980s and 1990s on 'images of ageing' in the mass media within contemporary consumer culture and the significance of these in everyday practice, continue to be cited, drawn on and challenged by other researchers (e.g. Kaufmann 1986; Gilleard and Higgs 2000; Öberg 2003; Ballard *et al.* 2005; Wainwright and B. Turner 2006).

Featherstone and Hepworth's studies in the 1980s and 1990s addressed the position of ageing within a postmodernist frame of reference where they detected an increasing tendency to extend the middle-age lifespan and lifestyle for as long as possible to put off the inevitability of deep old age. They also drew attention to what they termed the 'mask of ageing' (Featherstone and Hepworth 1991) whereby 'ageing' in postmodern culture may be increasingly understood as a mask which covers over or disguises the ways in which individuals actually feel about themselves. That is, there is a gap between the recognition of the outward physical appearance of the ageing body on the one hand and on the other, the subjective sense of age that individuals feel themselves to be inside (Kaufmann 1986; Öberg 2003). So, although individuals recognize the signs of ageing on their body, they may feel much younger inside. In

I Don't Feel Old (Thompson *et al.* 1990), which collated and analysed oral history testimonies on the participants' everyday experiences of later life, a 66-year-old man who would have been labelled as chronically sick in medical terms, remarked that: 'You don't feel any different than what you did when you were younger. You don't walk about saying, oh I'm old, I'm old, I feel old. You just feel the same . . .' (119).

The perceived signs of ageing are not individual, but are part and parcel of a restrictive discourse of ageing which, as Featherstone and Hepworth (1991) argue, impedes the possibilities for different modes of self-expression. In so doing, their approach presented a challenge to the limited vocabulary of ageing which restricted the potential for older people to express their ideas and sense of self 'outside of the prevailing stereotypes of the elderly' (Faircloth 2003: 17).

The problematic notions which surround the idea of old people continuing to be sexually active and their (seldom seen) unclothed bodies may be a case in point (Twigg 2000; Öberg 2003). These are founded on the taken for granted assumption that human beings naturally grow into sexuality at a certain point in adolescence and then grow out of it in later life. For example, dress, as Julia Twigg (2007) notes, is a significant medium in the surveillance and control of older bodies. Dress performs a key role in the expression of sexuality and Twigg argues that 'the exclusion of older women from fashion discourse largely arose from assumptions about sexuality' (295). In a recent study which explored how respondents assessed age in older people, Helle Rexbey and Jørgen Povlsen (2007: 61) found that the biological indicators of 'skin, eyes and hair colour' were most significant and that these were 'supplemented by perceptions of vigour, style, and grooming, and related to accepted codes of appearance'. Whilst the authors recognize that age categorization is a complex issue, the remarkably similar ways in which the respondents processed and assessed the age of one older person in comparison with another indicated that their interpretations were not simply subjective but involved shared cultural codes. Dress code was found to be important to *how* the respondents viewed or judged the images of older people, as age appropriate or not, particularly those of women. The term which is frequently used to describe older women who are deemed to be dressed inappropriately for their age is 'mutton dressed as lamb'. This term has a negative, gender-specific inference, which has a long history going back to the eighteenth century. It refers to older women behaving or dressing younger than the cultural codes of the particular period permit and it is related to ideas of morality. The term 'growing old gracefully, by contrast, is a positive term that refers to older women in particular who dress 'appropriately' for their age (see Fairhurst 1998).

Contemporary consumer culture facilitates the idea of choice and voluntarism in attitudes towards dress and to a certain extent, age is no longer viewed as a barrier to older women wearing fashionable clothes designed for a more youthful market. However, the participants in Rexbey and Povlsen's study remarked if older people crossed what they perceived to be the appropriate dress code line, particularly if the person in the photograph was dressed younger than their age. The majority of respondents reacted negatively to the image of one older woman with dyed black hair who was dressed in jeans, a low-cut top, high heels and gold jewellery, putting her into the age-old mutton dressed as lamb category. Participants in Eileen Fairhurst's earlier study (1998) responded in the same vein to a woman who was similarly dressed. However, the respondents in Rexbey and Povlsen's 2007 study assessed the woman as over 10 years younger in age than her chronological age (73 years). They also estimated that she was much younger than another woman of 73 years who was seen to be dressing appropriately for her age and thus, ageing gracefully. As the authors point out, it could be argued that this particular woman was successful in her attempt to look younger than she actually was at the time, if that indeed was her intention. Nevertheless, while fitting into the mutton dressed as lamb category, the woman's dress style was also described as 'cheap' by one respondent, which historically has connotations with prostitution. The same respondent considered that the woman was trying too hard to appear to be what she is not (young) and as indicated, the mutton dressed as lamb signifier also refers to moral decay, in this case through the attempt to deceive (ibid.). As Twigg (2007: 297) notes, the chastisement associated with the mutton dressed as lamb label hints at something more dangerous, which underlies its meaning, which is dressing 'inappropriately in a sexual way'. The results of Rexbey and Povlsen's study suggest that despite the cultural shifts indicated above in regard to a loosening of attitudes of what constitutes appropriate dress for older people, 'former patterns of age-ordering in dress still apply' (80) and that sexuality in older people, particularly women, continues to present a challenge to the moral (natural) order of things. Whilst for example, young women may purposefully 'play' with dressing 'tartily' in a postmodern fashion, it appears that this option is not readily open to older women.

While sexual activity in older people, particularly women, may be considered inappropriate in certain contexts, Barbara Marshall and Stephen Katz (2002) note that the traditional assumption that sexual decline is either a natural feature of old age or that sexual activity lies within the realm of youth and mid life and is less healthy or less seemly in later life, no longer holds to such an extent. However, they suggest that the relationship between age and sexuality remains problematic. Drawing on Foucault's approach to sexuality (1984), the authors examine the historical shifts in discourses on sexuality, ageing and the 'accelerated medicalization and technologization of sexual function' (Marshall and Katz 2002: 44) in recent years. In so doing, they argue that the decline or loss of male sexual function in later life is increasingly viewed as 'dysfunctional' or pathological. The introduction of the pharmaceutical remedy, Viagra, to assist penile erection, was hailed as a 'sexual revolution in itself'. According to Marshall and Katz, such technological interventions have contributed to the idea that older men can remain 'forever functional' sexually (ibid.), with the result that 'the ageing male is now a sexualized subject within a modern lifecourse regime' (63). The authors also note that increasingly Viagra is being marketed to younger men. This may be a sign that for men, too, age is moving younger, although they may not be so susceptible as women to what Kathleen Woodward (2006: 166) calls 'the youthful structure of the look' in her critical assessment of the ways in which the older female body is represented in 'visual mass culture' (162). Woodward additionally examines the work of three female artists, such as Louise Bourgeois, whom, she argues, 'expose, subvert, and exceed' the pressure on 'women to pass for younger once they have reached a "certain age" ' by showing the older female body as 'the continuing site of gender and sexuality' (ibid.).

Faircloth (2003: 17) also argues that fixing the aged in one spot disregards 'the different places in life they might actually place themselves'. Rather, he suggests, '[t]hey are not just old; they are many things'. Featherstone and Hepworth (1991), furthermore, perceive that the mask of ageing is not unchanging. They recognize contemporary generational shifts towards the idea of, for example, 'positive ageing' from the 1960s onwards which helped to open up possibilities for varied expressions of ageing from the perspective of the ageing person and how the aged are perceived. However, the positive images of older people, particularly third-agers, as presented in magazines and advertisements as healthy, active, financially independent (from child care and mortgages) and the freedom to pursue a renewed youthful lifestyle, have a downside (Featherstone and Hepworth 1995). Positive ageing, too, can become a kind of tyranny in as much as these affirmative consumer culture images of ageing carry normative implications. The process of ageing, in effect, becomes viewed as something which is 'endlessly open to construction or reconstruction . . . a matter of advances in medicine or in social reconstruction' (46). In a consumer-oriented society profoundly antagonistic to decay and dependency, the question that arises is what remains after this extended period of mid life which is 'typified by the imagery of positive ageing as period of youthfulness and active consumer lifestyles' (ibid.)?

Cultural studies, as Andrew Blaikie (1999) argues, like the sociology of ageing, has not paid much attention to the ageing body. However, by dint of the fact that studies have primarily addressed minority group issues in contemporary culture, and youth cultures, in particular, he considers that cultural studies' approaches have much to offer by examining older people in this light. The lack of attention paid to ageing from sociology and cultural studies in particular is all the more surprising in an era of rapid demographic change which has been taking place over the twentieth century and particularly since the 1960s. Blaikie's insightful monograph, *Ageing and Popular Culture* (ibid.) considers 'the cultural implications of continued population ageing', by means of a rigorous sociological–historical lens which draws on cultural studies' analyses of visual and popular culture, and is shot through with a commitment to the sociology of knowledge. He examines the ways in which 'the expanding popular leisure phase' within consumer culture is contributing to a narrowing of the gap between 'mid and later life'.

The general message from the literature on ageing which incorporates a social or cultural perspective, regardless of disciplinary base, is that the lives of older people today, at least as far as the third age is concerned, are dramatically different to those of their parents' generation. There is a perception that the generational gaps between older and younger people are becoming more blurred. Whilst the gap between the third age and youth may be 'moving downward' in certain respects, as Twigg (2007) indicates, where does that leave those who are beyond that stage in the life course or those who will enter into it in due course?

Thus, there is increasing evidence of a growing concern by those who are age-categorized as the 'young old', ably promoted and sanctioned by consumer culture, medical science and social and public policy alike, to stave off what can only be described as the descent into the following period of 'deep old age' which stretches before them, and which is characterized by the withdrawal from the mainstream of social life into the realm of invisibility and the inevitable end of life. While the hype around the new 'young old' or third agers is aptly captured in the media by the oft heralded slogan that nowadays '60 is the new 40', the very old, who are reaching the end of the life course, are associated with images of sickness and decay (Vincent 2003). However, it is important to note that these constructions are by no means uniform within individual societies in the present day or in their past histories, any more than they are across cultural divides.

THE PROBLEMS OF AGEING IN AN AGEING SOCIETY

In their 2002 policy report, the World Health Organization (WHO) noted that '[t]he proportion of people age 60 and over is growing faster than any other age group' in the 'developed' areas of the world and to a lesser extent, on a global scale (http://whqlibdoc.who.int/hq/2002/who_nmh_nph_02.8.pdf) (6). The growth in the ageing population is due to a number of factors over the twentieth century, such as dramatically decreasing birth rates, lower mortality rates and greater longevity through better living conditions, health care, disease control and diet. This is despite the fact that there have been 'setbacks' in age expectancy in other parts of the world such as Africa as a result of AIDS and in some new independent states, through the increase in 'cardiovascular disease and violence' (ibid.). The 'greying of the population' across the globe is expected to continue into the future. Currently, Europe has the largest proportion of people aged 60 and above. By 2025, it is expected that around one-third of the population will be over 60 years. The WHO considers that population ageing constitutes 'one of humanity's greatest triumphs' and at the same time, 'one of our greatest challenges' (1). Not only is the population getting older, the over 60s are also ageing, with about 1 per cent of the population in the developed areas aged 80 years and over, which is the 'fastest growing segment of the older population' (9).

Closer to home, the UK can be also be described as an ageing population. According to the Office of National Statistics' publication in 2009, the percentage of the population in the UK over the age of 65 increased from 15 per cent to 16 per cent from 1961 to 2009, while the population under 16 fell from 21 per cent to 19 per cent (http://www.cardi.ie/

publications/ageingfastestincreaseinthe%E2%80%98oldestold%E2%80%99). Over the same period, the fastest population growth has been in the 85 years and over age range; the 'oldest old' (ibid.). By 2034, it is expected that 23 per cent of the population will be over 65 years and over in comparison with 18 per cent under 16 years. Since 1984, the number of people reaching the age of 85 years and over has more than doubled; in 2009, this age cohort reached 1.4 million. It is projected that the number of people in the oldest old category will account for 5 per cent of the UK population by 2035. Increasingly, particularly since the onslaught of the economic downturn, it appears that the economic and social issues surrounding the care of an ageing population have taken on a status akin to that of a 'moral panic' (Cohen 1980).

However, concerns associated with an ageing society are not entirely new, as suggested above, although they may have been framed in different ways before. Population ageing has been rumbling as a potential 'problem' since the late-nineteenth century and it was at the turn of the twentieth century that gerontology began to develop in response to some of the concerns that were being raised around that time. Prior to this, in the late-eighteenth and early-nineteenth century, as Katz (1996) shows, there emerged in France the development of a new medical science discourse around the ageing body, 'senescence'. As a consequence, of the discursive shifts in medical knowledge, as Foucault (1973: 41) argues, death became reconfigured as a feature '*in* life', rather than '*of* life' (my emphasis) and in so doing, medical science's study of ageing 'became separate from the earlier treatises that focused on the promise of longevity'.

In *The Birth of the Clinic* (1973) Foucault documents the changes in perceptions and organizational systems in the field of medical knowledge at the end of the eighteenth century which eventually led to the formation of a new medical model. The old classificatory model of a 'medicine of the species' came to be replaced with a 'medicine of tissues' or 'anatomo-clinical medicine' in which the formation of the individual or, more precisely, 'the body' of the individual, 'as an object of scientific medical examination, and analysis' is revealed (Smart 1985: 26–9). Death, in this new model, effectively became a good position to analyse life, disease and pathology. In the eighteenth century hierarchical classificatory scheme, by contrast, death was considered to be the end of disease and the human body was treated as 'merely an object or space in which disease may be present' (28). Foucault maintains that it was within medical discourse that the individual (body) first became the 'object of positive knowledge' and that the idea of human beings as both subject and object of knowledge began to gain currency. Thus, for Foucault (1977: 191), medicine, as 'the first scientific discourse concerning the individual', paved the way for the subsequent development of the human sciences such as psychology and anthropology. Moreover, as John Vincent notes, these shifts in medical discourse not only contributed to a changed understanding of old age but also to the rise of the 'modern domination of medical definitions of the phenomenon' (2006: 682). Although social gerontology began to develop early in the twentieth century, it did not develop apace, as discussed below, until after the Second World War.

THE FIELD OF AGEING

Gerontology is a multi-disciplinary field which includes biological, social, psychological, anthropological, historical and cultural approaches. It has been accused of being characterized by 'the rhetoric of holism' by which the various methods of the 'hard' and 'soft' sciences are viewed as somehow coming together around the central concern of old age (Blaikie 1999: 11). In order to give authority to what is a relatively young field of study, researchers have often sought to utilize the theories and methods of the hard sciences to justify their findings. The consequence of this, as Blaikie indicates, 'has led to a "scientisation" of theory and methods in general, while social gerontology has been a poor relation to medical and psychological aspects of investigation' (ibid.). Thus, as Vincent (2003) argues convincingly, most of the research on ageing has been conducted in the areas of biology and medicine, as evidenced by, for example, the number of personnel involved in research and the concentration of funding directed towards these areas.

Despite the fact that there is 'no universally accepted theory of biological ageing' (33), there are two particular images that encapsulate, albeit in a very basic way, the key scientific theories of ageing. The first, according to Vincent, is the 'wear and tear' approach which is founded on the idea that the body wears out. The second is the 'time clock' approach, which is based on the notion that there is a certain point in the biological life cycle which 'triggers' the ageing process. The former is associated with 'advances in cellular ageing', while the latter is associated with 'developments in genetics' (133). Both approaches may be evidenced in relation to recent scientific endeavours to increase the lifespan of human beings beyond the current optimum of 120 years via on the one hand, the possibility of limiting cellular ageing and on the other, of limiting or 'manipulating genetic inheritance' (see also Seale 1998: 35–6).

Bio-gerontology is awash with debate on the state of knowledge in the area, on the validity of 'anti-ageing' medicine and the pros and cons of the scientific pursuit of extending the human lifespan, as Vincent (2006) and others (Higgs and Jones 2009; Lafontaine 2009) demonstrate. Vincent, invoking a cultural analytic paradigm, argues that 'human biological ageing needs to be understood through the social and environmental contexts in which it takes place' (2003: 134). In other words, ageing is not simply a biological fact, as the master narrative of medical discourse dictates; rather it is mediated through a range of social, cultural and environmental factors and conditions. At the same time, Vincent does not adopt a radical constructionist approach to the ageing body such as that proposed by Margaret Gullette (1997), for example. Gullette considers that the notion of 'entering' into old age is founded on the presumption that the body is in a determining state of physical decline, which is bound up with a range of cultural, historical and often tacit, assumptions and 'prescriptions' as what that stage *is*. She argues that it is a cultural trope which is founded 'on accepting the positivist claim of age ideology: that there is a real category of being there, separable from earlier stages or age classes and discontinuous from continuous processes as well' (159). As such, she argues that we need to resist the dominant 'decline narrative' of old age, whereby people are 'aged by culture'. However, Hepworth (2003: 98) suggests that concern with the 'ageing body/aged by culture' question in Gullette's work does not mean that 'biological ageing is marginalised as merely a socially constructed cultural artefact', as she includes examples of her own bodily problems in her work. Hepworth perceives that in Gullette's constructionist analysis: 'It is not, therefore, that biological decline in old age is a figment of the western imagination; it is the connections we make that are fictions and as such the conscious/unconscious reflections of a dominant ideology' (99).

Hepworth considers that Gullette not only critiques the dominance of the decline narrative, she also points out that there are other possible alternative discourses on ageing to choose from, which have emerged from 'literary culture' in recent years. He suggests that she is arguing for 'a more positive integration of the biological, psychological and social processes of ageing' (105). In this way, the experience of growing older involves 'a cultivation of a continuous life-story' that concerns the self moving towards the future, rather than focusing on 'an identity that has been buried somewhere in the past with a body of a previous identity' (ibid.); the youthful body/self, in other words. It should also be noted that Featherstone and Hepworth's socio-cultural approach to ageing discussed earlier does not refute that the ageing process involves biological change, although the dualistic split invoked by the mask of ageing, could be conceived as a harking back to a younger self (see Schwaiger 2006 on this), as opposed to one that encompasses a 'continuous life-story'.

Although ageing and death are recognized as universal features of human life, ageing among human beings is considered to be highly variable. For example, some functioning parts or organs of an individual's body may age more quickly than others. Professional dancers may present a good case in point where their extensive use of knees and hip rotation over long periods of time from a young age may give rise to 'ageing' joints long before they would 'normally' be expected to suffer from osteoarthritis or to be candidates for hip or knee replacement surgery. Further, as Carol Rambo Ronai's (2000: 285) field study of 'ageing' table dancers in their late teens and early twenties demonstrates, the experience of ageing is socially produced by individuals and 'through contexts that assign meaning to the physical body'. Ronai had eight years of insider access to the 'table dancing scene' in the area where she lived in the US. Prior to conducting her fieldwork on the everyday world of table dancers, Ronai was employed as a table dancer and as the research developed she became a 'complete member-researcher' (279). A table dancer of 25 years can be viewed as 'elderly' by customers, her employer and younger dancers in terms of a lessening of youthful bodily attractiveness, suppleness and 'sexual utility' and therefore in that particular milieu, earning potential. Ronai's study shows how certain young, 'elderly dancers' manage the ageing process within the context of their 'career pathway and occupational transitions' as striptease artists. She concludes that: '[d]ancers and others' stories and accounts suggest that ageing is an experience not necessarily just of later life, nor socially automatic and inevitable' (ibid.). As such, Ronai argues that age has to be separated from ageing in age research in order to uncover the ways in which an ageing experience is managed within given contexts. Similarly, Steven Wainwright and Bryan Turner's qualitative study of ballet dancers from the UK's Royal Ballet Company (2006), which involved interviewing dancers 'about age and embodiment', shows that even young, elite professional dancers in their early twenties talk about the fact that they can feel their bodies 'tightening up' and ageing, although this would not be noticeable in performance from an audience's viewpoint. Whilst the authors draw on Featherstone and Hepworth's (1991) notion of the 'mask of ageing', the results of their interviews suggest that there may be value in offering a variation of that concept. As the young ballet dancers *feel* much older than they look (the reversal of the mask of ageing), Wainwright and Turner suggest that the 'mask of youthful ageing' may be more appropriate in this context.

Although we can expect bodily functioning to slow down with advancing age, it is also the case that, as Vincent demonstrates, 'such processes can be controlled and retarded by human intervention' (2003: 134). Vincent cites the success of the anti-smoking campaign as an example whereby large numbers of regular smokers gave up the habit for good and as a consequence, the very real dangers to their health caused by their previous smoking habit decreased over time. Returning to the more specialized case of professional dancers, it is possible that by taking remedial action and re-training the body through alternative body techniques such as Pilates, for example, dancers can potentially lower the risk of being injured, which currently is around the 80 per cent rate in the UK (Laws 2005). Research has shown that injury is largely a consequence of over-use and strain as opposed to trauma (Solomon and Micheli 1986; Arendt and Kerschbaumer 2003; Bronner *et al.* 2003). As such, remedial action may reduce the wear and tear on the joints and other parts of the body. Moreover, as Wainwright and Turner's study illustrates, as dancers age, it takes longer to recover from injury and the body becomes more recalcitrant (Foster 1998).

Social gerontology, as Gubrium and Holstein (2000: 1) point out, draws on research from a range of disciplines such as 'anthropology, economics, history, nursing, political science, social psychology and social work, and sociology'. In part, this is due to the breadth of the research topics it covers, from, for example, 'interpersonal relationships, living arrangements and retirement, to social inequality, the politics of age, health, care giving, death and bereavement' (ibid.). The rapid development of social gerontology post 1945, according to the authors, was a response to the expanding population of older people and the vast increase of the birth rate after the Second World War, which was coined the 'baby boom'. The baby boomer age cohort has had particular relevance for social gerontology in the US and more recently in the context of Europe also (Phillipson 2007: 7).

The first cohorts of baby boomers were in their teens in the 1960s and are now in their 60s, forming what has been called the 'young old', 'third agers' and in a recent report, the end stages of the 'mid-lifers' or the 'sandwich generation' (Demey *et al.* 2011). They came of age in the 1960s, an era that witnessed the triumph of youth culture as the 'social value par excellence', the invocation of 'personal freedom and autonomy as the ultimate of individual accomplishment' (Lafontaine 2009: 58), and important shifts in cultural politics which led to the sexual revolution of the 1960s and 1970s. The baby boomers, it is argued, carried the values of youth with them as they grew older and are perceived as influencing and promoting changing attitudes to ageing downwards, leading the way for new products and goods circulating in the market place in the 1980s which promoted the idea of 'maturity as a second youth' (ibid.).

However, categorizing the baby boomers as an age cohort gives rise to several problems. To begin with, the idea that the boomers represent a 'meaningful social category', with which people in this age cohort identify, does not take account of the fact that this 'will vary greatly from country to country' (Phillipson 2007: 7). The surge in birth rates in the industrial countries after the end of the Second World War was in fact not even. In Britain, for example, there were two separate surges in the birth rates, one in 1947 and the other in 1964, while the increase in the US and Australia took place over a longer period of time, from the middle of the 1940s to the middle of the 1960s (ibid.). Finland, on the other hand, had a more condensed rise in birth rates, which followed in the wake of the soldiers returning home and which ended in the early 1950s (ibid.). Research conducted in Finland suggests that boomers, like their counterparts in the US, display a stronger collective identification with the term, which appears to cut across social differences (Biggs *et al.* 2007). Research in the UK, however, based on interviews with people from different social groups who were born between 1945 and 1954, suggests that this age cohort does not have a strong identification with the term boomer (ibid.). Rather, the findings indicate that this cohort has a stronger sense of linkage with the notion of the '60s generation'. The results further show that consumption is important for this age cohort, particularly 'in the areas such as clothing and bodily maintenance' (Phillipson 2007: 9), which it may be suggested, sets this cohort apart from pre-war age cohorts.

A further criticism of the boomer cohort approach is that it assumes that boomers constitute one single grouping, which, in turn, denies difference in class, gender and race within and across different countries. Chris Gilleard and Paul Higgs (2007) argue that a generational approach of the 'third age' is a much stronger conceptual sociological framework for addressing the shifts in ideas and attitudes towards ageing than that of an age cohort. '[U]nderstanding the role of the sixties' cultural revolution for the emergence of the third age', the authors argue, 'offers a broader conceptual understanding of the transformation of later life than that provided by the more restrictive and restricting framework of a baby boom cohort' (13). Moreover, they suggest that the strong self-identity with the term baby boomers in the US has less to do with reflecting an 'age cohort structuring influence' on the consciousness of individuals and more to do with the role 'the market and the media play in shaping their social identities' (ibid.). However, it is interesting to note that the term 'baby boomer' has been given much attention in the UK mass media, particularly since the onslaught of the global economic downturn in 2008 and is the subject of much political debate.

The baby boomers are coming up to retirement and given their numbers and the fact that people are now living longer, they will inevitably place an ever-increasing burden on the state over the next twenty to thirty years. This has become an increasing source of concern in the politics of age. Critics argue that the baby boomers, the 'never had it so good' generation, the original 'me' generation that invented youth culture and has never grown up, have robbed their children of their future by their profligate lifestyle, lack of saving and investment for future generations, in stark contrast to their forebears. There has been a recent flurry of popular books on this subject (e.g. Howker and Malik 2010; Willetts 2010; Beckett 2011). The boomers, critics suggest, have contributed to the huge burden of debt that now rests securely on the shoulders of youth, the very idea of which they invented and which they have now disinherited, through their irresponsible behaviour. This 'intergenerational unfairness', it is argued, will inevitably lead to a war between the older and younger generations over the next twenty years as the boomers age (http://www.independent.co.uk/news/uk/politics/will-the-babyboomers-bankrupt-britain-1936027.html). However, this standpoint assumes that most baby boomers are wealthy and over-privileged, but as the *Independent* article by Paul Vallely cited above reported in 2010, 'Many of the population, including the majority of

Baby-Boom pensioners, live on less than £22,000 a year' (ibid.). A lot of boomers have to live off their state pension and not, as is sometimes stated, a final salary pension to see them comfortably through their old age. Many also have difficulty in keeping warm over the winter months. Thus, '[f]or many Baby Boomers the greatest inequality is not between the generations. It is within the existing one' (ibid.). Moreover, it is not as if 'population panics' over the rapid expansion or the decrease in birth rates at particular periods in the history of modern industrial society are something new (Blaikie 1999). Consider, for example, Robert Malthus' concern with population growth in the nineteenth century which was first expressed in his highly controversial yet influential treatise, *Essay on the Principle of Population* (1798), in which he set out the consequences of unchecked population growth on increasingly scarce resources, such as food, as a result of having many more mouths to feed.

As suggested before, divisions between different stages of the life course are not fixed social categories. The concept of childhood, for example, was 'invented' by the Victorians and restrictions on child labour were also put into the legal system over that era, which eventually led to the barring of children from the workforce. The idea of adolescence came into being in the first decade of the twentieth century and as Blaikie (1999) demonstrates, this stage is clearly identifiable in contemporary post Second World War notions of youth culture. The idea of a retirement phase at a fixed age was also firmly established in the early twentieth century in modern industrial societies such as the US and Western Europe. Although civil servants were the first group of workers to have a set retirement age of 65 in the UK, state pensions were not introduced until 1908 for workers over 70 years. Retirement represented a withdrawal from the workforce and a marker for defining someone as old. The requirement to retire at the age of 65 for men and 60 years for women was made a condition in 1948. From that point on, 'age is defined through retirement and thus to employment' (Hearn 1995: 100). The withdrawal from active social engagement with work was also accompanied by negative stereotypes of old age, which was equated with dependency, isolation and eventual senility, or 'the problems of the elderly' (Blaikie 1999: 51). In response to the fact that people are now living longer, recent changes were made to the legal state pensionable age in the UK. From 2010, women's state pensionable age was increased from 60 to 65 years for those born on or after the first of April 1950. From 2018, the state retirement age for both men and women will rise to 66 in the first instance, with proposals for further increases over time (http://www.direct.gov.uk/en/Pensionsandretirementplanning/StatePension/DG_4017919).

Historical research suggests that before the nineteenth century people worked until they could no longer manage it and only then were they defined as being old (Vincent 2003:10). Prior to industrialization, old age was 'individual and flexible', which, by implication, suggests a more positive state. However, as Featherstone and Hepworth (1993: 261) argue, it is important not to romanticize old age in the past, because there is also sufficient evidence to show that power and wealth were crucial to how old people were defined, and that 'the stigmatisation of old age was *in practice* a familiar feature of everyday life' [my emphasis] in the pre-industrial period.

As indicated above, gerontology took a social problems approach in the 1960s, focusing on social welfare and policy issues associated with older people. Some social gerontologists in this period advanced the idea of growing old in terms of a theory of 'mutual disengagement' between the ageing individual and society, while others adopted a counter theory based on 'activity'. Disengagement theory was based on observations that as people age, they gradually withdraw from the activities and people they engaged with in middle age. The disengagement from the social system was considered to be a 'natural' response to the disengagement from the work and social roles in mainstream society (Cumming and Henry 1961). Ageing successfully from this perspective entails 'the acceptance and desire for a process of disengagement from social life' (Havighurst 1961: 8) and the restoration of an equilibrium between the ageing individual and society which was present in middle age, but which was temporarily destabilized as a consequence of the readjustment from one role in middle age to the other in old age. Disengagement theory prompted a great deal of research, debate and controversy, leading to modifications of the theory and arguments against its static, functionalist perspective which took no account of changing social roles and presented negative stereotypes of old age (Maddox 1964; Youmans 1969).

By contrast, 'activity theory', was based on the presumption that people should continue those 'activities and attitudes of middle age for as long as possible' (Havighurst 1961: 8) and that they should substitute these for others once they are no longer able continue with them or have to give them up. Activity theorists argued that the older person's withdrawal from mainstream society is a consequence of society's disengagement from the ageing person, 'against the will and desire of the person' (9). From the activity theory perspective, successful ageing involves staying active and not withdrawing from social life and in effect, keeping busy. This theory, too, was criticized for being based on an equilibrium model of society and as such, it was assumed that when 'change occurred the typical response was to restore the previous equilibrium' (Atchley 1989: 183). It was also argued that activity theory privileged individual adaptation over and above larger structural concerns such as differences between older people founded on class, race or gender (Katz 2000). Despite these criticisms, according to Katz, the idea of activity embedded in activity theory has had a lasting legacy in the field of gerontology in regard to the 'problem of adjustment'. This is particularly evidenced in positive notions of ageing, as discussed earlier. However, several critics have pointed to links between positive ageing approaches that stress activity and 'consumerist ideologies' (ibid.). David Ekerdt (1986: 239) argues that the emphasis on activity and keeping busy in retirement 'is morally managed and legitimated on a day-to-day basis in part by an ethic that esteems leisure that is earnest, occupied, and filled with activity–a "busy ethic"'. The busy ethic draws its authority from the work ethic, according to Ekerdt and in so doing, 'it legitimates the leisure of retirement' and at the same time, provides a 'definition for the retirement role' (243).

A third theory of ageing which sought to overcome the limitations of disengagement theory is the 'continuity theory of normal ageing', which emerged in the 1970s (Atchley 1989). It is premised on the idea that people in middle age and later life have a preference to carry with them their beliefs, patterns of behaviour, likes and dislikes, friendships and way of life as they are making adaptive changes to their lives, both internally (psychological) and externally (social). The process of adaptation is one of continuity rather than disjuncture (Atchley 1971). This approach sets out a distinction between 'normal' and 'pathological' ageing based on the 'adaptive principles that people who are normally ageing could be expected to follow' (Atchley 1989: 184). As it is based on evolution as opposed to stasis, it is recognized that what counts as normal will vary from culture to culture, and it may be assumed will change over time. Continuity theory has been criticized for its marked distinction between normal and pathological ageing which excludes those older people who have disabilities or are chronically ill (Putman 2002) and for its '"male model" of life satisfaction' (Calasanti 1996: S18).

In the 1970s, hard on the heels of the development of the 'greying population' as a cultural and then, a social movement with political organizational teeth (see Featherstone and Hepworth 1995), the social study of ageing shifted its emphasis towards 'approaches which question assumptions about ageing being made by those who are non-aged' (Blaikie 1999: 12). It is during this period that terms such as 'ageist', and positive and successful ageing gain ascendency. Thus, according to Blaikie (ibid.), there has been a growing interest in social gerontology in 'the social construction of ageing, in lived experience' and in 'normal as against pathological ageing'. In part, these shifts contributed to a clear divide being drawn between bio-gerontology, which is firmly situated under the umbrella of biology, and social gerontology, which leans more towards a sociological perspective. Nonetheless, as Blaikie indicates, sociologists are prone to point out that despite the shift towards constructivism and an increasing attention to understanding the everyday lives of older people, gerontology tends to focus on social policy issues rather than addressing sociological problems: 'Pensions, older people, and retirement represent objective social problems, rather than subjects of investigation in their own right, which is how a sociological approach would proceed' (ibid.).

The disciplinary basis of sociology and gerontology can be traced back historically. The labels researchers use to 'classify' old people are constructs that emerge from their particular disciplinary base. Thus, as Katz (1996) points out, social gerontology exists within a disciplinary framework through which the social construction of ageing is formulated, which was founded on the medical model. Using a Foucauldian approach, Katz argues that 'discourses on ageing are also disciplinary in the sense that they contribute to the processes of regulation and control of old age' (Blaikie 1999: 12). Discourses on ageing are not fixed in stone from this perspective; they change over time and they have ramifications for people's lives as they are translated into action on ageing. For Higgs *et al.* (2009), for example, the 'will to health' has become a key discourse in the ageing agenda which impacts significantly on the lives of older people and their engagement with the health agenda. The will to health underpins the imperative of maintaining the appearance of a 'productive' old age, after exiting from the system of economic production, which somewhat echoes Ekerdt's (1986) idea of 'busy ethic', mentioned earlier. Here a distinction between 'natural ageing' and 'normal ageing' is proposed. The construct of natural ageing is associated with an acceptance that the body will decay and decline with age, while normal ageing embraces the idea of trying to remain 'forever young' (Schwaiger 2006) from mid life and on through the third age to delay the descent into dependency, disability, disease and finally, death, which is characterized by the fourth age (Baltes and Smith 2003). Higgs *et al.* (2009: 687) examine the ways in which these discourses are linked to ideas of anti-ageing in the context of the changes in consumerist society after the 'golden age of welfare capitalism', in which the values of youth already evident in the 1960s, along with increased individuation and lifestyle choices (Giddens 1991) in the face of prospective 'risks' (Beck 1992) have become embedded and 'viewed as guiding principles of social action' (Higgs *et al.* 2009: 688).

Later life today, as suggested before, 'is very different in character from that experienced by previous generations' (ibid.). Ageing and the hazards of growing older become a challenge to the imperative of the 'autonomous self' where active agents are expected to make rational choices and act accordingly. Foucauldian inspired governmentality scholars like Nikolas Rose (1996) have noted the ways in which new technologies have sought to improve the management of the autonomous self or what Rose (2001) terms, 'somatic individuality'. Higgs *et al.* (2009: 689) are concerned to explore how 'the reconstruction of contemporary later life' has impacted upon and been enacted by the first generation to go through these changes, which has led to the techniques and responsibilities of the autonomous self becoming 'part of that generational habitus'. The authors consider that the ideas around health, aided and abetted by anti-ageing medicine and products, are key to the relentless search for an active, productive third age lifestyle, founded on the distinction between natural and normal ageing.

The relationship between activity and well-being in older life appears to be central to the contemporary discourse on ageing in gerontology, health promotion, the leisure industry and everyday opinion (Katz 2000). Activity is seen as contributing to a productive, normal ageing lifestyle on the one hand and on the other, facilitating the management of risk factors such as illness and disability and loneliness which may come with old age, which were perceived to be the lot of older people prior to the tropes of positive and productive ageing coming into vogue. Body maintenance through physical activity is viewed as a keystone of a healthy lifestyle and care of the self across the life course in health promotion and is particularly important in an ageing society (Paulson 2005; Kreutz

2008). It features strongly in positive and productive ageing and may be translated into what Katz sees as the regulatory control of later life through the emphasis on 'busy bodies'. It is significant that three of the books discussed in this chapter which were published between 1999 and 2003 have images on the front cover of older white women, actively engaging in physical pursuits, as opposed to being slumped in a chair looking disengaged or appearing to be frail. On the cover of *Ageing and Popular Culture* (Blaikie 1999) is a photograph of three older women in swimsuits standing in a single file in the sea or swimming baths, with the water up to chest height. With outstretched arms and raised chins, they seem to be concentrating hard on getting their positions right in what might be either synchronized swimming or water aerobics. The photograph is part of a series taken by Georgina Ravenscroft in the mid 1990s, depicting older people on holiday in seaside resorts such as Bournemouth, Madeira and Tenerife (see Blaikie 1999: 166). The cover of Vincent's book, *Old Age* (2003), depicts a social tea dance event which foregrounds an older woman laughing, with her body in full motion and skirt a twirling as her partner swings her round by her arm in a vigorous jive dance. The photograph was taken by David Tothill and is entitled *Elderly Couple Dancing at a Tea Dance* (http://www.photofusion.org). The cover of Faircloth's edited collection, *Ageing Bodies* (2003), has one large studio-based, 'aesthetic' soft focus image and two smaller detailed images of the same shot. The large image portrays the back view of a well-toned, slim, white haired older woman in a white, sleeveless leotard with low-cut back and white tights seated on the floor in a difficult spinal twist yoga pose, which stretches the spine and lifts the neck and head upwards following the direction of the diagonal pull of the upper body. I know it is difficult because I tried to imitate the pose and it clearly requires a strong sense of body centring and attention to line, as well as flexibility and strength. The smaller images show her outstretched, well-arched right foot and slim ankle in one and in the other, the lower right arm and hand appears to be pushing with strength into the floor behind the body. The photograph and a number of others by Ryan McVey portraying the same woman in various poses which are entitled, *Elderly Woman in a Yoga Pose*, are to be found in Getty Images (http://www.gettyimages.co.uk).

Although the images focus on older women engaging in different activities, nevertheless they portray older bodies of various shapes and sizes in action, which attest to the positive and productive ageing, 'will to health' discourse, which older people are encouraged to buy into and many do so.[3] The activities in question also require certain bodily skill sets and control, and not only the woman in the yoga pose, as already noted.

The idea of physical activity as an antidote to the vagaries of old age is not entirely new either, although it may have taken on a different (more positive) significance in contemporary culture. As Katz (2000) notes, in the 1950s the concern to remain active in the post-retirement period was raised partly in response to the negative aspects of disengagement theory, as described earlier. Many studies emphasize the health and psychological benefits to be gained through physical culture activity in later life (see Marquez *et al.* 2009), the rationale for which is situated within a social policy agenda for an ageing society.

The following section considers the role of physical activity in later life by drawing and reflecting on a research project conducted in 2000–2001 by a small research team of Lesley Cooper and Helen Thomas. The theoretical and methodological framing was situated within a socio-cultural framework which sought to understand the meanings that dancing had for participants in the study, rather than operating on the basis of a social policy agenda (Thomas and Cooper 2003). The research was mentioned briefly in regard to the discussion on ethnographic rapport in Chapter 4. In the context of this chapter the title is particularly apt: *Dancing into the Third Age: Social Dance as Cultural Text*.[4]

DANCING INTO THE THIRD AGE; DANCING, AGEING AND EMBODIMENT

Dancing into the Third Age aimed to assess the meanings of social dance as a cultural practice for groups of participants who were entering or had entered the third age. The research centred on people over the age of 60 years who participated in dance events in the two inner city south-east London boroughs of Southwark and Lewisham, and the county of Essex to the north-east of London. This enabled the researchers to generate a map of the range of dance events and workshops which older people regularly attended in the two inner-city areas and the suburban and rural milieu of north Essex. In total, 24 sites were visited initially and, of these, 11 sites representing the breadth of dance-type activities which older people engaged in were visited again in order to go into more depth. It became abundantly clear early on in the project that there were a surprising number and range of regular dance events involving people in the older age ranges. In effect, our older participants were travelling across London and Essex several times a week in buses, tubes, trains and cars, with their dance shoes in bags, to engage in their preferred dancing activities. Some preferred to stick with one type of dancing, while others liked to take part in different dance genres.[5]

The linking of dance and older people in these contrasting social environments was quite unusual in dance studies or cultural studies at this time. The history of social dancing from the 1960s has been largely associated with specific youth subcultures and their identification with popular music, although the emphasis was and still is to a large extent, overwhelmingly focused on the music, with the dancing being given short shrift (Ward 1993, 1997). Further, although representations of ageing had begun to attract the attention of cultural and social analysts, particularly in regard to the young old, negative stereotyping of older people as 'the elderly' was still pervasive in the mass media and popular discourse (Featherstone and Hepworth 1995; Gubrium and Holstein 2000). Thus, the project sought to bring the shadowy presence of dance and old age into the sights and (sites) of social and cultural analysis and dance studies. Research has expanded somewhat over the past ten or so years, with the majority of the work in this area focusing on the physiological and psychological impact of dance on older people's well-being in health-based studies, as evidenced in Connolly and Redding's 2010 review of the literature on dance and well-being (http://www.trinitylaban.ac.uk/media/315435/literature%20review%20impact%20of%20dance%20elderly%20populations%20final%20draft%20with%20logos.pdf). There remains little systematic research in social and cultural analysis, with Wainwright and Turner's study of ballet dancers (2003, 2004 and 2006) and Elizabeth Schwaiger's (2012) study of older professional dancers being the exceptions rather than the rule.

THEORETICAL FRAMING

The study was located within three cultural and academic milieux; the anthropology and sociology of dance, the social, cultural and academic framing of ageing, and literature on the body which had burgeoned in the 1990s, as discussed in Chapter 1. Equal weighting was given to the social contexts of the dance events encountered by the researchers, the significance of dancing in the daily life of the participants, and the aesthetic components of the dance forms practised. The anthropology/sociology of dance (e.g. Browning 1995; Savigliano 1995; Buckland 1999) provided the basis of the ethnographic approach that was adopted in the study. The lack of attention to older people in dance and cultural studies was counterbalanced by a review of the literature on ageing, especially where it related to the experiences of old age and the cultural and social framing of the older person (Thompson *et al.* 1990; Gubrium and Holstein 2000, 2003). This revealed a tension between, on the one hand, a growing concern with the active 'youthful' healthy ageing body and the reality of the invisibility of the older people in everyday life on the other (see Thomas and Miller 1997).

The sociology of the body, unsurprisingly, was central to the project. In order to overcome the problems with constructivism on the one hand and biological essentialism on the other, the researchers drew on approaches that emphasized the lived experience of the body (e.g. Csordas 1993, 1994; Featherstone *et al.* 1991; Crossley 1995), and particularly the construct of embodiment, which insists that consciousness and experience are habitually embodied and intertwined. As noted in Chapter 1, research on the body was criticized for being overly theoretical at the expense of empirical analysis and little attention had been paid to the moving body. Thomas Csordas' concept of 'somatic modes of attention' (1993), which brings together insights from Merleau-Ponty's theory of embodiment with Bourdieu's model of habitus was particularly pertinent in this regard. Somatic modes of attention are 'culturally elaborated ways of attending to and with one's body in surroundings that include the embodied presence of others' (138). As such, the construct contains the notion of both a culturally and socially lived body and one that has its own agency. Embodiment and somatic modes of attention, from this perspective were important to both our understanding of dance, conceived as a situated embodied aesthetic practice (Thomas 2003) and as already discussed, to ageing (B. Turner 1995). It also fitted well with constructionist approaches in dance ethnography where dance is viewed as a form of embodied cultural knowledge (Ness 1992; Bull 1997; Sklar 2000, 2001). In so doing, this approach, I now suggest with hindsight, situated the project in relation to ideas discussed already in this chapter in regard to linking the doing of age and ageing within the tripartite relation between 'the society, the body and the self' (Frank 1996: 53).

METHODOLOGY/RESEARCH METHODS

An ethnographic approach involving participant observation and semi-structured and unstructured or 'ethnographic' interviews (Brewer, 1994) was employed to gain an in-depth understanding of both the dancers and the dancing. Thirty-one in-depth interviews were conducted, coded and transcribed. The ages of respondents ranged from 60–89 years, with the majority between 70–79 years. Importantly, we not only observed the dancing but also participated in it. The researchers themselves, within this approach, become part of the study and cannot be viewed as 'objective' observers. Thus, the age, race, gender and personality of the researchers built into and coloured the research. Our engagement in the dancing and our dancing skills, or the lack of them, as indicated in the previous chapter, fed into the way that the research developed and generally provided 'insider' access. However, the move from outsider to insider was not fixed. Rather, in reality, the researchers oscillated between the two frames of reference, as perhaps sociologist Georg Simmel suggested in his classic methodological essay, 'The Stranger' (1950 [1908]). Both researchers were a generation below the participants in the study and from their perspective we appeared to inhabit a very different social world. They generally thought that we were younger than our chronological age, which links back to Laz's articles on performing age (1998, 2003). While the influence of class, gender and race, was very much part of the researchers' background coming through from the 1960s and 1970s, we could also hear the influence of that era of cultural politics in the talk of some of the participants. At the same time, there were moments when we felt we were stepping into a different bygone world, particularly in relation to the formal customs and manners in the context of social dancing for example. Similarly, the respondents often wondered what we would do when we got older as our generation did not know how to dance properly (see Thomas 2000 on this point). All of this, of course, was before *Strictly Come Dancing* on BBC 1 television gave ballroom and Latin partner dancing a much needed shot in the arm, although there had been evidence in the late 1990s that younger people were beginning to show an interest in competition dancing (Thomas and Miller 1997).

The research included video recording the full range of dance forms practised; out of a total of 24 dance sites visited, nine were filmed by the researchers. These sites were one modern sequence dance club, a tap dance class and a Scottish country dance club in Essex and in south east London, a creative dance class, one Darby and Joan club tea dance catering for mostly people in the fourth age range, two mixed social dance venues, an Irish pensioners' tea dance, and an African Caribbean quadrille workshop. The edited tapes were taken back to four sites so that the participants could comment on their interpretations and understanding of their performances. The researchers were given dance videos taken by members in two of the dance sites some years before, which portrayed some of our participants dancing. Some of that older footage was inserted into the edited videos of the two groups. The video material constituted a valuable resource for analysing the movement styles and the skills the dancers employed in specific contexts. One of the videos showed that some of the participants' dancing skills had visibly improved with age since the first video was taken ten years before, which contrasts sharply with the decline model discussed before. Emmanuelle Tulle's research on elite veteran runners (2008) also found the runners' running improved in quite complex ways as they aged and that they saw themselves as athletes, not as a shell of their former athletic self.

The following section briefly outlines two aspects from the overall findings that relate to notions about the importance of dance activity in later life from the participants' perspective in this study, which included among other aspects: aesthetics of dance and ageing, feeling good, having fun, social engagement, dance and music, keeping fit and young, dancing skill, dance and the ageing body and mind, and death and bereavement. Here I will briefly discuss the aesthetics of dance and the ageing body, and the ageing body and mind.

DANCE AESTHETICS AND THE AGEING BODY

The first thing to note is that is that aesthetic aspects of dancing were important to the majority of the men and women in the study, whether it was a class or a workshop although it was particularly notable in social dancing. Social dancing was mostly modern sequence dancing or mixed ballroom dancing along with set dancing. Looking aesthetically pleasing on the dance floor is an important aspect of all good social dancing, according to most of the men and women respondents. The form of dress and style of appearance depended on the venue and the dance style involved. While the level of smartness and formality of dress varied in different social dance settings and occasions, the correct shoes, with non-slip soles, were essential for the majority of social dancers. Respondents differed in their understanding of what counts as a good dancer depending on the style of dance performed. Social dancers especially spent a lot of time discussing the intricacies of what it means to be a good dancer. When sequence and ballroom dancers spoke about someone they considered to be a good dancer or what good dancing meant to them, they invariably included the look of the moving dancing body in time and space. In Scottish country dancing, by contrast, a relaxed upright carriage with arms hanging down to the sides were important attributes of the dance aesthetic. Whilst it was recognized that age does affect the capability of dancers, younger respondents usually made allowances for the ageing body when commenting on who was a good dancer; often pointing to someone much older than themselves who they considered to be a good dancer. There was an implicit assumption that age is not always relevant when it comes to good dancing and that dancers can overcome increasing disabilities through their own skill and grace. One respondent, for example, spoke of 'an old girl who is 86 and who can do anything that anybody does'. Two men in their 40s had dancing partners who were about 30 years older than them who displayed considerably more dancing skills, too.

Good dancers, we suggested, can challenge the negative construction of the ageing body as one that is undesirable, and un-aesthetically pleasing, by creating an agreeable picture on the dance floor and can increase the performers' cultural capital. On reflection, this seems to concur with the notion of 'growing old gracefully' with its attendant implications as mentioned earlier, and indeed, this was the title of an article based on some of the findings of the project (Cooper and Thomas 2002). However, it should also be noted that Cooper wanted to call the paper 'growing old disgracefully' to challenge the implicit demand for normative compliance in the former term, which was also evidenced in the research by those who refused to conform to the more conventional images of older people, by dressing flamboyantly for example.

DANCE AND THE AGEING BODY AND MIND

Despite the advanced age of some of the dancers, most appeared to be fit and energetic on the dance floor. Even those with serious disabilities did not seem disabled when they were dancing; they appeared both skilled and full of life and vigour. The older dancers frequently acknowledged that even those who hobble in the streets can and do get up and dance as if there is nothing wrong with them. Our observations of individuals in their late 70s and 80s on the dance floor bore this out consistently. Dancers may have had cancer, or other serious illnesses, or may have had an operation only weeks before, and yet on the dance floor they appeared to be fit, energetic and able.

All respondents believed that dancing wards off the ailments of old age especially the stiffening up of the body and the atrophy of the mind. One modern sequence dancer, for example, had a serious operation a year before and was told by his surgeon that his fitness was due to his dancing and was largely responsible for his rapid recovery.

But the research also found that dancing can highlight the failing body as well as the active, vital body. While learning steps and formations, as in the quadrille for instance, can help the practitioners to keep the ageing body flexible and engage the memory, it can reveal that older people learn more slowly, are not so agile and forget more quickly than when they were younger. We witnessed regularly the difficulty that some of the older dancers had with memorizing steps and placements in the quadrille, while younger dancers were more able to pick these up and recall them from week to week. But this could have been because some of the older participants were new to the style of dancing involved in the quadrille, which has complicated floor patterns, rather than an issue of ageing. The video the researches were given by the teacher showed some of the older quadrille participants performing a group partner dance in a flowing sensuous understated style that demonstrated a different range of aesthetic bodily skills from their often stilted and unsure performance style when dancing the quadrille. However, even some modern sequence dancers who were very familiar with the genre commented on the difficulties they experienced remembering the ever-increasing number of new dances they had to learn.

The aim of the research was to explore and convey the importance and the multifarious benefits of dance for a significant proportion of the older people in the study. It was found that social dancing provided not just the opportunity for physical exercise and for increased fitness, but a uniquely positive experience for the participants, at a time when so much of the literature on ageing focused on loss and negativity. What struck attention in all of the research sites was the fact that the dancers loved to dance and to socialize with others. The analysis found that dance can mitigate the public invisibility of the older person and it can show that old age is not such a fixed phenomenon. It can bring both a real and mythic experience of aliveness, fitness, fun and flexibility, along with a sense of continuity when it is most needed. Part of the specific meaning that dance brings is that it can (re) generate a feeling of communitas, which has also been noted in other studies on social dance (e.g. Cowan 1990; O'Connor 1997). Thus, the research drew on continuity theory as well as positive ageing. Although not discussed here, while the research pursued an anti-decline approach to the third age and explored the failing body within that, it nevertheless treated the fourth age as a period of decline. Moreover the study was more focused on a particular age group, rather than ageing per se, although this was not obvious at the time.

CONCLUSION

This chapter has sought to elucidate the ways in which older bodies have been constructed and perceived in contemporary culture and in studies of ageing from different disciplines and perspectives. It started from the premise, following Laz (2003), that age may be viewed as performance, a doing word, in which the body is centrally located in relation to the self and society. The meanings associated with age and ageing are not fixed in stone but change over time and space. These shifts were addressed by recourse to a consideration of the ways in which perceptions of ageing and age have shifted, particularly over the twentieth century, which witnessed the growth of an ageing population and which brought with it advantages and myriad potential social problems. The chapter considered how these changes have impacted on representations of older people within a consumer-oriented culture that privileges youthfulness and the social analysis of ageing and age. The idea of a new third age (Laslett 1989), which emphasized a new lease of life based on activity rather than decline, was explored through a discussion and critique of a study of ageing and dancing conducted in 2000–2001, which adopted a socio-cultural approach in contrast to the policy or health-oriented research on the benefits of dancing to health in late life. It was suggested the promise of the third age was evidenced in the behaviour and talk of the dancing participants and on reflection, in the researchers' theoretical and methodological framing as well.

CONCLUSION

This book started off by considering the entrance of 'the body' into the domain of social and cultural analysis which, over a period of time, captured the intellectual imagination of a generation of scholars to such an extent that the very idea of the 'body project' (Shilling 1993) became a project and an object of analysis, in and of itself, in many different areas of research. The oft cited idea that the body is generally out of awareness (Hall 1969; Goffman 1972) or 'absent' (Leder 1990) as we go about our daily business until something calls it into attention was found not to be quite as uniform as has sometimes been suggested. It was proposed that our enabling and recalcitrant bodies act on our sense of the self and others in ways that make it difficult to segregate the body and the self in the context of the social world in which we move and live out our lives. Directing close attention to the body *in* everyday life as a mechanism for considering the body *and* everyday life would have been one way to proceed. I chose to address theories, themes and concerns through the lens of a range of performance practice case studies where bodily actions/notions loom large through the 'not-everyday' but which speak to and of the everyday in nuanced ways. Performance practices often emerge through the vernacular and/or draw consummately from the intricacies of the everyday and in so doing, speak to/of it in challenging ways. The aim was to cast some light on the complexities of trying to grapple with the slippery, multifarious notions of the body or bodies in (and) everyday life. As John Law (2004: 2) points out, the subjects that social science deals with are more often than not 'complex, diffuse and messy' and when it then tries to describe such matters in a definitive, simplified manner, it usually 'makes a mess of it'.

It is not for me to say if the journey I decided to take has succeeded to any extent in opening up a small window of research possibilities. While recognizing that the considerable attention paid to the body in social and cultural analysis over the past twenty or so years has impacted critically on more traditional theoretical or methodological approaches, there is still a lot of room for manoeuvre and further critical analysis. It is perhaps a vain hope that this book will raise some questions and contribute to the further opening up of the 'body problem' through the small looking glass of the performance practices in question. There has been an increasing interest in sociology in performance in recent years. It is important for researchers to be aware that there are traditions of thought in other disciplinary areas outside of sociology that have addressed themes such as performance and performativity in a more sustained manner. In many ways, I admire Norman Denzin (1997: 123) for celebrating performance ('texts') and performance studies in ethnography, which for him 'has crossed that luminal space that separates the scholarly text from the performance'. It may be that social science researchers can learn through critical engagement with this 'performance' work that predates the current interest and knowledge in the field, as I also attempted to demonstrate in the chapter on cultural and social performances, in order not to reinvent the wheel and advance the knowledge base in a more open, fluid and productive way. In the final chapter where I reflected on the *Dancing into the Third Age* study, it became clear to me that the publications that emerged out of the project, while novel in certain respects, presented a rather fixed and uniform view of the participants in the study, despite our best attempts to avoid this. As such, it may mean that we should consider the possibilities of 'abandoning method', as David Phillips (1973) once proposed, which is hard for sociologists steeped in methodological rigour to so do.

Conclusions usually describe the contents of the book in detail and present a summing up of the progress of the book's journey to its end point to close the discussion down. This book, for me at least, has represented an opening rather than a closing inasmuch that doing the research for the chapter on boxing, in particular, led me to my next research project – a comparative analysis of boxing and dancing with a strong ethnographic streak!

NOTES

CHAPTER 2

1. Although there was an interest in the notion of performance in psychology in regard to roles, this book does not specifically focus on psychological approaches.
2. Carlson (2004) provides a broad-based survey of uses of the terms 'performance' and 'performativity', which I discovered after I had written much of this chapter. Carlson approaches his survey from a performance studies perspective which includes a more detailed discussion of areas such as linguistics, psychology and performance studies than I am able to do here. A further text in theatre studies which is very useful as it covers a lot of ground is Simon Shepherd and Mick Wallis, *Drama/Theatre/Performance* (2004).

CHAPTER 5

3. It is also telling that these images were immediately recognizable to me, and presumably other readers, too, as portraying older women in particular physical activities which, many can and do participate in for pleasure within the organized framework of health and fitness clubs and social centres, or at least the first two popular culture type activities. My first reading of the photos was confirmed after tracing the images back to their respective source via the Internet and finding the accompanying titles of two of them, as indicated above.
4. This research project was funded in 2000–2001 by the then Arts and Humanities Research Board (AHRB), now the Arts and Humanities

Research Council (AHRC), The AHRB provided further match funding in 2002 to finalize the report and to set up a website www.dance.gold.ac.uk to give details of the project and to edit the video material of nine of the dance sites down to three minutes each. The website is still available but the videos are no longer accessible on the website and can be found in the University of the Arts London Research Online website (ualresearchonline.arts.ac.uk).

5 *Dancing into the Third Age* built on an earlier qualitative study of a ballroom dance hall in south-east London (Thomas and Miller 1997; Thomas 2000), where my interest in studying dancing and ageing was first raised. Much of the discussion in this section is drawn directly and quite liberally from the project's website cited above and the larger report published in 2003 by the researchers for copyright reasons and in order to not re-invent the wheel. See Thomas and Cooper (2003) for a detailed discussion of the methods used to establish the extent and range of available dance activities and the process of mapping and coding which emerged from observing the events. See also Cooper and Thomas (2002) on the similarities and differences between the tea dances in London and modern sequence dances in Essex.

BIBLIOGRAPHY

Aaron, M. (ed.) (2001) *The Body's Perilous Pleasures*, Edinburgh: Edinburgh University Press.

Adams, P. (1996) *The Emptiness of the Image*, London: Routledge.

Amaya, H. (2004) 'Photography as technology of the self: Matuscha's art of breast cancer', *International Journal of Qualitative Studies in Education*, 17, 4: 563–73.

Anderson, N. (1923) *The Hobo: The Study of the Homeless Man*, Chicago: Chicago University Press.

Arendt, Y. and Kerschbaumer, F. (2003) 'Injury and overuse pattern in professional ballet dancers', *Zeitschrift für Orthopädie und ihre Grenzgebiete*, 141: 349–56.

Atchley, R. C. (1971) 'Retirement and leisure: participation: continuity or crisis?', *The Gerontologist*, 11, 1: 13–17.

—— (1989) 'A continuity theory of normal aging', *The Gerontologist*, 29, 2: 83–90.

Atkinson, P. (1990) *The Ethnographic Imagination*, London: Routledge.

Atwood, M. (1991) 'The female body', in L. Goldstein (ed.) *The Female Body: Figures, Styles, Speculations*, Ann Arbor: University of Michigan Press.

Austin, J. L. (1962) *How to Do Things with Words: The William James Lectures Delivered at Harvard 1955*, London: Clarendon Press.

Back, L. (2004) 'Inscriptions of Love', in H. Thomas and J. Ahmed (eds) *Cultural Bodies: Ethnography and Theory*, Oxford: Blackwell.

Bakhtin, M. M. (1984 [1968]) *Rabelais and His World*, Bloomington: Indiana University Press.

Ballard, K., Elston, M. A. and Gabe, J. (2005) 'Beyond the mask: women's experiences of public and private ageing during midlife and their use of age-resisting activities', *Health: An Interdisciplinary Journal for the Social Study of Health, Illness and Medicine*, 9, 2: 169–87.

Baltes, P. B. and Smith, J. (2003) 'New frontiers in the future of aging: from successful aging of the young old to the dilemmas of the fourth age', *Gerontology*, 49, 2: 123–35.

Banes, S. (1980) *Terpsichore in Sneakers: Post-Modern Dance*, Boston: Houghton Mifflin Company.

Barba, E. (1991) 'Theatre anthropology', in E. Barba and N. A. Savarese *Dictionary of Theatre Anthropology: The Secret Art of the Performer*, London: Routledge.

Barrett, M. (1988) *Women's Oppression Today: The Marxist/Feminist Encounter*, London: Verso.

Barthes, R. (1977) *Image-Music-Text*, Glasgow: William Collins Sons and Co.

—— (1985) *The Grain of the Voice Interviews 1962–1980*, New York: Hill and Wang.

Bateson, G. and Mead, M. (1942) *Balinese Character: A Photographic Analysis*, New York: Special Publications of the New York Academy of Sciences.

Bauman, R. (1992) 'Performance', in R. Bauman (ed.) *Folklore, Cultural Performances and Popular Entertainment*, Oxford: Oxford University Press.

Beck, U. (1992) *Risk Societies: Towards a New Modernity*, London: Sage Publications.

Becker, H. S., Greer, B., Hughes, E. C. and Strauss, A. L. (1961) *Boys in White: Student Culture in Medical School*, Chicago: Chicago University Press.

Beckett, F. (2011) *What Did the Baby Boomers Ever Do for Us?*, London: BiteBack Publishers.

Bell, D., Caplan, P. and Karim, W. J. (eds) (1993) *Gendered Fields: Women, Men and Ethnography*, London: Routledge.

Bennett, A. (2000) *Popular Music and Youth Culture: Music, Identity and Place*, Basingstoke: Macmillan Press.

Benthall, J. and Polhemus, T. (eds) (1975) *The Body as a Medium of Expression*, London: Allen Lane.

Benton, T. (1991) 'Biology and social science: why the return of the repressed should be given a (cautious) welcome', *Sociology*, 25, 1: 1–29.

Betterton, R. (ed.) (1987) *Looking On: Images of Femininity in the Visual Arts and Media*, London: Pandora Press.

—— (1996) *Intimate Distance*, London: Routledge.

Bhavani, K. K. and Coulson, M. (1986) 'Transforming socialist-feminism: the challenge of racism', *Feminist Review*, 23, June: 81–92.

Biggs, S., Phillipson, C., Rebecca, L. and Money, A.-M. (2007) 'The mature imagination and consumption strategies: age and generation in the development of a United Kingdom baby boomer identity', *International Journal of Ageing and Later Life*, 2, 2: 31–59.

Birdwhistell, R. (1973) *Kinesics in Context: Essays in Body-Motion Communication*, Harmondsworth: Penguin University Books.

Birringer, J. (1993) *Theatre, Theory, Postmodernsim*, Bloomington: Indiana University Press.

Blacking, J. (ed.) (1977) *The Anthropology of the Body*, London: Academic Press.

Blaikie, A. (1999) *Ageing and Popular Culture*, Cambridge: Cambridge University Press.

Blum, V. (2003) *Flesh Wounds: The Culture of Cosmetic Surgery*, Basingstoke: Palgrave Macmillan.

Bordo, S. (1993) *Unbearable Weight: Feminism, Western Culture and the Body*, Berkeley: University of California Press.

—— (1994) 'Reading the male body', in I. Goldstein (ed.) *The Male Body: Features, Destinies, Exposures*, Ann Arbor: University of Michigan Press.

Bourdieu, P. (1977) *Outline of a Theory of Practice*, Cambridge: Cambridge University Press.

—— (1983) 'Erving Goffman: discoverer of the infinitely small', *Theory, Culture and Society*, 2, 1: 112–13.

—— (1984) *Distinction: A Social Critique of the Judgement of Taste*, London: Routledge & Kegan Paul.

—— (1990a) *In Other Words: Essays Towards a Reflexive Sociology*, Cambridge: Polity Press.

—— (1990b) 'Programme for a sociology of sport', in P. Bourdieu *In Other Words: Essays Towards a Reflexive Sociology*, Cambridge: Polity Press.

—— (1993) *Sociology in Question*, London: Sage Publications.

—— (1996) *The Rules of Art*, Cambridge: Polity Press.

—— (2003) 'Participant objectivation', *Journal of the Royal Anthropological Institute*, 9, 2: 281–94.

Bourdieu, P. and Wacquant, L. J. D. (1992) *An Invitation to Reflexive Sociology*, Chicago: Chicago University Press.

Boyne, R. and Rattansi, A. (eds) (1990) *Postmodernism and Society*, London: Macmillan Press.

Brah, A. (1996) *Cartographies of Diaspora: Contesting Identities*, London: Routledge.

Brennan, T. (1989) *Between Psychoanalysis and Feminism*, London: Routledge.

Brewer, J. D. (1994) 'The ethnographic critique of ethnography: sectarianism in the RUC', *Sociology*, 28, 1: 231–44.

Bronner, S., Ojofeitimi, S. and Rose, D. (2003) 'Injuries in a modern dance company: effect of comprehensive management on injury incidence and time loss', *American Journal of Sports Medicine*, 31, 3: 365–73.

Browning, B. (1995) *Samba: Resistance in Motion*, Bloomington: Indiana University Press.

Buckland, T. J. (ed.) (1999) *Dance in the Field: Theory, Methods and Issues in Dance Ethnography*, Basingstoke, Macmillan Press.

Bull, M. and Back, L. (eds) (2003) *The Auditory Culture Reader*, Oxford: Berg.

Bull, C. J. C. (a.k.a. Cynthia Novack) (1997) 'Sense, meaning, and perception in three dance cultures', in J. C. Desmond (ed.) *Meaning in Motion: New Cultural Studies of Dance*, Durham and London: Duke University Press.

Bulmer, M. (1983) 'The methodology of the taxi-dance hall: an early account of Chicago ethnography from the 1920's', *Urban Life*, 12, 1: 95–102.

—— (1984) *The Chicago School of Sociology: Institutionalization, Diversity and the Rise of Sociological Research*, Chicago: University of Chicago Press.

Burke, K. (1945) *A Grammar of Motives*, Cleveland: Meridan.

Burkitt, I. (1999) *Bodies of Thought: Embodiment, Identity and Modernity*, London: Sage Publications.

Burns, T. (1992) *Erving Goffman*, London: Routledge.

Burt, R. (1995) *The Male Dancer: Bodies, Spectacle and Sexuality*, London: Routledge.

—— (1998) *Alien Bodies: Representations of Modernity, "Race" and Nation in Early Modern Dance*, London: Routledge.

Butler, J. (1990) *Gender Trouble: Feminism and the Subversion of Identity*, London, Routledge.

—— (1993) *Bodies That Matter: On the Discursive Limits of "Sex"*, London: Routledge.

—— (1994) 'Gender as performance: an interview with Judith Butler'. Interview by Peter Osborne and Lynne Segal, London, 1993. *Radical Philosophy*, 67: 32–39.

—— (1999) 'Perfomativity's social magic', in R. Shusterman (ed.) *Bourdieu: A Critical Reader*. Oxford: Blackwell Publishers.

Calasanti, T. M. (1996) 'Gender and life satisfaction in retirement: an assessment of the male model', *Journal of Gerontology: Social Sciences*, 51B: S18–S29.

Caplan, J. (ed.) (2000) *Written on the Body: The Tattoo in European and American History*, London: Reaktion Books.

Caplan, P. (1988) 'Engendering knowledge: the politics of ethnography' (Pt.I), *Anthropology Today*, 14, 4: 8–12.

Carby, H. (1982) 'White woman listen! Black feminism and the boundaries of sisterhood', in The Contemporary Centre for Cultural Studies (ed.) *The Empire Strikes Back: Race and Racism in 70s Britain*, London: Hutchinson.

Carlson, M. (2004) *Performance: A Critical Introduction*, London: Routledge.

Caute, D. (1972) *The Illusion*, New York: Harper Row.

Chaney, D. (1994) *The Cultural Turn: Scene-Setting Essays in Contemporary Cultural History*, London: Routledge.

Chernin, K. (1986) *The Hungry Self: Women, Eating and Identity*, London: Virago Press.

Chomsky, N. (1965) *Aspects of the Theory of Syntax*, Cambridge, MA: MIT Press.

Clifford, J. (1986) 'Introduction: Partial Truths', in J. Clifford and G.E. Marcus (eds) *Writing Culture: The Poetics and Politics of Ethnography*, Berkeley: University of California Press.

—— (1988) *The Predicament of Culture: Twentieth-Century Ethnography, Literature, and Art*, Cambridge, MA: Harvard University Press.

Clifford, J. and Marcus, G. E. (eds) (1986) *Writing Culture: The Poetics and Politics of Ethnography*, Berkeley: University of California Press.

Coffey, A. (1999) *The Ethnographic Self: Fieldwork and the Representation of Identity*, London: Sage Publications.

Cohen, S. (1980) *Folk Devils and Moral Panics: The Creation of the Mods and Rockers*, 2nd ed., Oxford: Martin Robinson.

Connell, R. W. (1995) *Masculinities*, Berkeley: University of California Press.

Cook, R., Segal, N. and Taylor, L. (eds) (2003) *Indeterminate Bodies*, Basingstoke: Palgrave Macmillan.

Cooper, L. and Thomas H. (2002) 'Growing old gracefully: social dance in the third age', *Ageing and Society*, 22, 6: 689–708.

Coupland, J. and Gwyn, R. (eds) (2003). *Discourse, the Body and Identity*, Basingstoke: Palgrave Macmillan.

Cowan, J. (1990) *Dance and the Body Politic in Northern Greece*, Princeton: Princeton University Press.

Coward, R. (1984) *Female Desire*, London: Paladin Grafton Books.

Coward, R. and Spence, J. (1986) 'Body talk: a dialogue between Ros Coward and Jo Spence', in P. Holland, J. Spence and S. Watney *Photography/Politics Two*, London: Comedia.

Cressey, P. G. (1968 [1932]) *The Taxi-Dance Hall: A Sociological Study in Commercialized Recreation and City Life*, New York: Greenwood Press.

—— (1983) 'A comparison of the roles of the "sociological stranger" and the "anonymous stranger" in field research', *Urban Life* 12, 1: 102–20.

Crossley, N. (1994) *The Politics of Subjectivity: Between Foucault and Merleau-Ponty*, Aldershot: Avebury Press.

—— (1995) 'Merleau-Ponty, the elusive body and carnal sociology', *Body & Society* 1, 1: 43–62.

—— (2001a) *The Social Body: Habit, Identity and Desire*, London: Sage Publications.

—— (2001b) 'Embodiment and social structure: a response to Howson and Inglis', *The Sociological Review*, 49, 3: 318–26.

—— (2006) 'In the gym: motives, meaning and moral careers', *Body & Society*, 12, 3: 23–50.

Csordas, T. J. (1993) 'Somatic modes of attention', *Cultural Anthropology*, 8, 2: 135–56.

—— (1994) 'Introduction: the body as representation and being-in-the-world', in T. J. Csordas (ed.) *Embodiment and Experience: The Existential Ground of Culture and Self*, Cambridge: Cambridge University Press.

—— (2002) *Body, Meaning, Healing*, Basingstoke: Palgrave Macmillan.

Cumming, E. and Henry, W. E. (1961) *Growing Old: The Process of Disengagement*, New York: Basic Books.

Cunningham-Burley, S. and Backett-Milburn, K. (eds) (2001) *Exploring the Body*, Basingstoke: Palgrave Macmillan.

Curti, L. (1998) *Female Stories, Female Bodies: Narrative, Identity and Representation*, Basingstoke: Macmillan Press.

Daly, A. (1987) 'The Balanchine woman: of hummingbirds and channel swimmers', *The Drama Review*, 31, 1: 8–21.

—— (1987/88) 'Classical ballet: a discourse of difference', *Women & Performance*, 3, 2: 57–66.

Daly, M. (1978) *Gyn/ecology: The Metaethics of Radical Feminism*, Boston: Beacon Press.

Davis, K. (1995) *Reshaping the Female Body: The Dilemma of Cosmetic Surgery*, New York: Routledge.

—— (ed.) (1997) *Embodying Practices: Feminist Perspectives on the Body*, London: Sage Publications.

—— (2003) *Dubious Equalities and Embodied Differences: Cultural Studies on Cosmetic Surgery*, Lanham: Rowman and Littlefield Publishers Inc.

de Beauvoir, S. (1972 [1949]) *The Second Sex*, Harmondsworth: Penguin Books.

de Lauretis, T. (1986) *Technologies of Gender: Essays on Theory, Film and Fiction*, Basingstoke: Macmillan Press.

de Saussure, F. (1974 [1915]) *Course in General Linguistics*, London: Fontana.

Debord, G. (2005 [1967]) *The Society of the Spectacle*, New York: Zone Books.

Deleuze, G. and Guattari, F. (1983) *Anti-Oedipus: Capitalism and Schizophrenia*, New York: Viking Press.

DeMello, M. (2000) *Bodies of Inscription: A Cutural History of the Modern Tattoo Community*, Durham: Duke University Press.

Demey, D., Berrington, A., Evandrou, M. and Falkingham, J. (2011) 'The changing demography of mid-life, from the 1980s to the 2000s', *Population Trends*, 145: 1–19.

Denzin, N. K. (1989) *The Research Act: A Theoretical Introduction to Sociological Methods*. Englewood Cliffs: Prentice Hall.

—— (1997) *Interpretive Ethnography:* London: Sage Publications.

Denzin, N. K. and Lincoln, Y. S. (eds) (1994) *Handbook of Qualitative Research*, 3rd ed., Thousand Oaks: Sage Publications.

Derrida, J. (2001 [1978]) *Writing and Difference*, London: Routledge.

Desmond, J. C. (ed.) (1997) *Meaning in Motion: New Cultural Studies of Dance*, Durham: Duke University Press.

Dingwall, R. (1997) 'Accounts, interviews and conversations', in G. Miller and R. Dingwall (eds) *Context and Method in Qualitative Research*. London: Sage Publications.

Doane, M. A. (1987) *The Desire to Desire: The Woman's Film of the 1940s*, London: Macmillan Press.

Douglas, M. (1970) *Purity and Danger: An Analysis of Concepts of Pollution and Taboo*, Harmondsworth: Penguin.

—— (1973) *Natural Symbols: Explorations in Cosmology*, Harmondsworth: Penguin.

—— (1975) *Implicit Meanings: Essays in Anthropology*, London: Routledge & Kegan Paul.

Doyle, J. and Roen, K. (2008) 'Surgery and embodiment: carving out subjects', *Body & Society*, 14, 1: 1–7.

Dubin, S. C. (1983) 'The moral continuum of deviancy research: Chicago sociologists and the dance hall'. *Urban Life*, 12, 1: 75–94.

Duncan, N. (ed.) (1996) *Bodyspace*, London: Routledge.

Durkheim, E. (1952 [1897]) *Suicide: A Study in Sociology*, London: Routledge & Kegan Paul.

—— (1964 [1895]) *The Rules of Sociological Method*, New York: Free Press.

—— (1968 [1912]) *The Elementary Forms of Religious Life*, London: George Allen & Unwin.

—— (1984 [1893]) *The Division of Labour in Society*, New York: The Free Press.

Early, G. (1994) *The Culture of Bruising: Essays on Prizefighting, Literature and Modern American Culture*, Hopewell, NJ: Ecco Press.

Ekerdt, D. J. (1986) 'The busy ethic: moral continuity between work and retirement', *The Gerontologist*, 26, 3: 239–44.

Ellis, C. and Bocher, A. P. (2000) 'Autoethnography, personal, narrative, reflexivity', in N. K. Denzin and Y. S. Lincoln (eds) *Handbook of Qualitative Research*, 2nd ed., Thousand Oaks: Sage Publications.

Entwistle, J. (2000) *The Fashioned Body: Fashion, Dress and Modern Social Theory*, Cambridge: Polity Press.

Entwistle, J. and Wilson, E. (eds) (2001) *Body Dressing*, Oxford: Berg.

Evans, M. and Lee, E. (eds) (2002) *Real Bodies: A Sociological Introduction*, Basingstoke: Palgrave Macmillan.

Faircloth, C. A. (2003) 'Introduction: different bodies and the paradox of ageing: locating aging bodies in images and everyday experiences', in C. A. Faircloth (ed.) *Ageing Bodies: Images and Everyday Experiences*, Walnut Creek and Oxford: Altamera Press.

Fairhurst, E. (1998) ' "Growing Old Gracefully" as opposed to "Mutton Dressed as Lamb": The social construction of recognising older women', in S. Nettleton and J. Watson (eds) *The Body in Everyday Life*, London: Routledge.

Falk, P. (1994) *The Consuming Body*, London: Sage Publications.

Featherstone, M. (1991) 'The body in consumer culture', in M. Featherstone, M. Hepworth and B. S. Turner (eds) *The Body: Social Processes and Cultural Theory*. London: Sage Publications.

—— (ed.) (2000) *Body Modification*, London: Sage Publications.

—— (2010) 'Body, image and affect in consumer culture', *Body & Society*, 16, 1: 193–221.

Featherstone, M. and Hepworth, M. (1984) 'Changing images of retirement: an analysis of representations of aging in the popular magazine "Retirement Choice" ', in D. B. Bromley (ed.) *Gerontology: Social and Behavioural Perspectives*, London: Croom Helm.

—— (1991) 'The mask of ageing and the postmodern lifecourse', in M. Featherstone, M. Hepworth and B. S. Turner (eds) *The Body: Social Processes and Cultural Theory*, London: Sage Publications.

—— (1993), 'Images of Ageing', in J. Bond and P. Coleman (eds) *Ageing and Society: An Introduction to Social Gerontology*, London: Sage.

—— (1995) 'Images of positive aging: a case study of "Retirement Choice" magazine', in M. Featherstone and A. Wernick (eds) *Images of Aging: Cultural Representations of Later Life*. London: Routledge.

Featherstone, M., Hepworth, M. and Turner, B. S. (eds) (1991) *The Body: Social Processes and Cultural Theory*, London: Sage Publications.

Featherstone, M. and Wernick, A. (eds) (1995) *Images of Aging: Cultural Representations of Later Life*, London: Routledge.

Fielding, N. (1994) 'Ethnography', in N. Gilbert (ed.) *Researching Social Life*, London: Sage Publications.

Firestone, S. (1970) *The Dialectic of Sex*, New York: Bantam Press.

Foster, S. L. (1996) *Corporealities: Dancing Knowledge, Culture and Power*, London: Routledge.

—— (1998) 'Choreographies of gender', *Signs: Journal of Women in Culture and Society*, 24, 1: 1–33.

Foucault, M. (1973) *The Birth of the Clinic*, London: Tavistock Press.

—— (1977) *Discipline and Punish: The Birth of the Prison*, Harmondsworth: Peregrine Books.

—— (1980) *Power/Knowledge: Selected Interviews and Other Writings 1972–1977*, London: Harvester Press.

—— (1984) *The History of Sexuality*, Harmondsworth: Peregrine Books.

—— (1986) 'Neitzche, genealogy, history', in P. Rabinow (ed.) *The Foucault Reader*, Harmondsworth: Penguin Books.

Frank, A. W. (1990) 'Bringing bodies back in: a decade review', *Theory, Culture and Society*, 7: 131–62.

—— (1991) 'For a sociology of the body: an analytical review', in M. Featherstone, M. Hepworth and B. S. Turner (eds) *The Body: Social Processes and Cultural Theory*, London: Sage Publications.

—— (1995) *The Wounded Storyteller: Body, Illness, and Ethics*, Chicago: University of Chicago Press.

—— (1996) 'Reconciliatory alchemy: bodies, narratives and power', *Body & Society*, 2, 3: 53–71.

Fraser, M. and Greco, M. (eds) (2005) *The Body: A Reader*, Oxford: Routledge.

Fraser, S. (2003) *Cosmetic Surgery, Gender and Culture*, Basingstoke: Palgrave Macmillan.

Friedan, B. (1992 [1962]) *The Feminine Mystique*, London: Penguin Books.

Frost, L. (2001) *Young Women and the Body: A Feminist Sociology*, Basingstoke: Palgrave Macmillan.

Gaines, J. and Herzog, C. (eds) (1990) *Fabrications: Costume and the Female Body*, London: Routledge.

Gamman, L. and Makinen, M. (1994) *Female Fetishism: A New Look*, London: Lawrence and Wishart.

Gamman, L. M. and Marshment, M. (eds) (1988) *The Female Gaze: Women as Viewers of Popular Culture*, London: The Women's Press.

Garfinkel, H. (1984 [1967]) *Studies in Ethnomethodology*, Cambridge: Polity Press.

Gatens, M. (1996) *Imaginary Bodies: Ethics, Power and Corporeality*, London: Routledge.

Geertz, C. (1974) 'From the native's point of view', *Bulletin of the American Academy of Arts and Sciences*, 28, 1: 27–45.

—— (1975) *The Interpretation of Cultures*, London: Hutchinson and Co.

—— (1983) *Local Knowledge*, New York: Basic Books.

Giddens, A. (1991) *Modernity and Self-Identity*, Cambridge and Oxford: Polity Press, in association with Blackwell Publishers.

Gilleard, C. and Higgs, P. (2000) *Cultures of Ageing: Self, Citizenship and the Body*, London: Prentice Hall.

—— (2007) 'The third age and the baby boomers: two approaches to the social structuring of later life', *International Journal of Ageing and Later Life* 2007, 2, 2: 13–20.

Gilman, S. (1995) *Health and Illness: Images of Difference*, London: Reaktion Books.

—— (1999) *Making the Body Beautiful: A Cultural History of Aesthetic Surgery*, Princeton: Princeton University Press.

Goellner, E. W. and Shea Murphy, J. (eds) (1995) *Bodies of the Text: Dance as Theory, Literature as Dance*, New Brunswick: Rutgers.

Goffman, E. (1963) *Behaviour in Public Places: Notes on the Social Organization of Public Gatherings*, New York: The Free Press.

—— (1971 [1959]) *The Presentation of Self in Everyday Life*, Harmondsworth: Penguin Books.

—— (1972) *Relations in Public: Microstudies of the Public Order*, Harmondsworth: Penguin Books.

—— (1979) *Gender Advertisements*, London: Macmillan Press.

Goldberg, R. (1979) *Performance: Live Art 1909 to the Present*, London: Thames and Hudson.

Goldstein, L. (ed.) (1991) *The Female Body: Figures, Styles, Speculations*, Ann Arbor: University of Michigan Press.

—— (ed.) (1994) *The Male Body: Features, Destinies, Exposure*, Ann Arbor: University of Michigan Press.

Griffin, S. (1978) *Women and Nature: The Roaring Inside Her*, New York: Harper & Row.

Grindon, L. (1996) ' "Body and Soul": The structure of meaning in the boxing film genre', *Cinema Journal*, 35, 4: 54–59.

Grosz, E. (1993) 'Bodies and knowledges: feminism and the crisis of reason', in A. Alcoff, L. and E. Potter (eds) *Feminist Epistemologies*, London: Routledge.

—— (1994) *Volatile Bodies: Towards a Corporeal Feminism*, Bloomington and Indianapolis: Indiana University Press.

—— (1995) *Space, Time, and Perversion*, New York: Routledge.

Grosz, E. and Probyn, E. (eds) (1995) *Sexy Bodies: The Strange Carnalities of Feminism*, London: Routledge.

Gubrium, J. F. (2000) 'Narrative practice and the inner worlds of the Alzheimer's disease experience', in P. J. Whitehouse, K. Maurer and J. F. Ballenger (eds) *Concepts of Alzheimer Disease*, Baltimore: Johns Hopkins University Press.

Gubrium, J. F. and Holstein, J. A. (eds) (2000) *Aging and Everyday Life*, Oxford: Blackwell Publishers.

—— (2003) 'The everyday visibility of the aging body', in C. A. Faircloth (ed.) *Aging Bodies: Images and Everyday Experiences*, Oxford: Altamira Press.

Gullette, M. M. (1997) *Declining to Decline: Cultural Politics of the Midlife*, Charlottesville: University of Virginia Press.

Hall, E. T. (1969) *The Hidden Dimension*, Garden City, New York: Anchor Books.

Hammersley, M. (1992) *What's Wrong with Ethnography*, London: Routledge.

Hammersley, M. and Atkinson, P. (1995) *Ethnography: Principles in Practice*, London: Routledge.

Haraway, D. J. (1991) *Simians, Cyborgs and Women: The Reinvention of Nature*, London: Free Association Books.

Harris, G. (1999) *Staging Femininities: Performance and Performativity*, Manchester: Manchester University Press.

Hastrup, K. (1992) 'Out of anthropology: the anthropologist as an object of dramatic representation', *Cultural Anthropology* 7, 3: 327–45.

—— (1995) *A Passage to Anthropology: Between Experience and Theory*, London: Routledge.

Harvey, D. (1989) *The Condition of Postmodernity*, Oxford: Basil Blackwell.

Hauser, T. (1991) *Mohammad Ali: His Life and Times*, New York: Simon Schuster.

Havighurst, R. J. (1961) 'Successful aging', *The Gerontologist*, 1, 1: 8–13.

Hearn, J. (1995) 'Imaging the ageing of men', in M. Featherstone and A. Wernick (eds) *Images of Ageing: Cultural Representations of Later Life*, London: Routledge.

Hebdige, D. (1979) *Subcultures: The Meaning of Style*, London: Methuen.

Heiskanen, B. (2006) 'On the ground and off: the theoretical practice of professional boxing', *European Journal of Cultural Studies*, 9, 4: 481–96.

Hepworth, M. (1995) 'Positive ageing: what is the message?', in R. Bunton, S. Nettleton and R. Burrows (eds) *The Sociology of Health Promotion; Critical Analysis of Consumption*, London: Routledge.

—— (2003) 'Ageing bodies: aged by culture', In J. Coupland and R. Gwyn (eds) *Discourse, the Body and Identity*, Basingstoke: Palgrave Macmillan.

Higgs, P. and Jones, I. R. (2009) *Medical Sociology and Old Age: Towards a Sociology of Health in Later Life*, London: Routledge.

Higgs, P., Leontowitsch, M., Stevenson, F. and Jones, I. R. (2009) 'Not just old and sick – the "will to health" in later life', *Ageing and Society*, 29, 5: 687–707.

Hirschhorn, M. (1996) 'Orlan: artist in the post-human age of mechanical reincarnation: body as ready (to be re-) made', in G. Pollock (ed.) *Generations and Geographies in the Visual Arts: Feminist Readings*, London: Routledge.

Hirst, P. Q. and Woolley, P. (1982) *Social Relations and Human Attributes*, London: Tavistock.

hooks, b. (1992) *Black Looks: Race and Representation*, Boston: South End Press.

—— (1994) *Outlaw Culture: Resisting Representations*, London: Routledge.

Howker, E. and Malik, S. (2010) *Jilted Generation: How Britain has Bankrupted its Youth*, London: Icon Books.

Howson, A. and Inglis, D. (2001) 'The body in sociology', *The Sociological Review*, 49, 3: 297–317.

Hughes, H. S. (1974) *Consciousness and Society*, St. Albans: Paladin.

Hughes-Freeland, F. (1998) 'Introduction', in F. Hughes-Freeland (ed.) *Ritual, Performance, Media*, London: Routledge.

Huyssen, A. (1986) *After the Great Divide: Modernism, Mass Culture and Postmodernism*, London: Macmillan Press.

Hymes, D. (1975) 'Breakthrough to performance', in D. Ben-Amos and K. S. Goldstein (eds) *Folklore: Performance and Communication*, The Hague: Mouton.

Ince, K. (2000) *Orlan: Millennial Female*, Oxford: Berg.

Inglis, D. (2005) 'The sociology of art: between cynicism and reflexivity', in D. Inglis and J. Hughson (eds) *The Sociology of Art: Ways of Seeing*, Basingstoke: Palgrave Macmillan.

Irigaray, L. (1985) *This Sex Which Is Not One*, Ithica: Cornell University Press.

Jay, M. (1999) 'Returning the gaze: the American response to the French critique of ocularcentrism', in G. Weiss and H. F. Faber (eds) *Perspectives on Embodiment: The Intersections of Nature and Culture*, London: Routledge.

Jenkins, R. (1992) *Pierre Bourdieu*, London: Routledge.

Johnson, C. (1994) 'A phenomenology of the black body', in L. Goldstein (ed.) *The Male Body: Features, Destinies, Exposures*, Ann Arbor: University of Michigan Press.

Johnston, L. (1998) 'Reading the sexed bodies and spaces of the gym', in H. J. Nast and S. Pile (eds) *Places through the Body*, London: Routledge.

Jones, M. (2008) 'Makeover culture's dark side: breasts, death and Lolo Ferrari', *Body & Society*, 14, 1: 89–104.

Kappeler, S. (1986) *The Pornography of Representation*, Cambridge: Polity Press.

Katz, S. (1996) *Disciplining Old Age: The Formulation of Gerontological Knowledge*, Charlottesville: University of Virginia Press.

—— (2000) 'Busy bodies: activity, aging and the management of everyday life', *Journal of Aging Studies*, 14, 2: 135–38.

Kaufmann, S. R. (1986) *The Ageless Self: Sources of Meaning in Later Life*, Madison: University of Wisconsin Press.

Khon, T. (2003) 'The Aikido body: expressions of group identities and self-discovery in martial arts training', in N. Dyck and E. P. Archetti (eds) *Sport, Dance and Embodied Identities*, Oxford: Berg.

Koritz, A. (1995) *Gendering Bodies/Performing Art: Dance and Literature in Early Twentieth-Century British Culture*, Ann Arbor: University of Michigan Press.

Krakow, A. (2005) *The Total Tattoo Book*, New York: Warner Brooks Inc.

Krauss, R. (1985) *The Originality of the Avant-Garde and Other Modernist Myths*, Cambridge, MA: MIT Press.

Kreutz, G. (2008) 'Does partnered dance promote health? The case of tango Argentina', *The Journal of the Royal Society for the Promotion of Health*, 128, 2: 79–84.

Kristeva, J. (1982) *Powers of Horror: An Essay on Abjection*, New York: Columbia University Press.

—— (1984) *Revolution and Poetic Language*, New York: Columbia University Press.

Kroeber, A. (1952) *The Nature of Culture*, Chicago: Chicago University Press.

Kroker, A. and Kroker, M. (1987) 'Panic sex in America', in A. Kroker and M. Kroker (eds) *Body Invaders: Sexuality and the Postmodern Condition*. Basingstoke: Macmillan Educational.

Kuppers, P. (2007) *The Scar of Visibility: Medical Performances and Contemporary Art*, Minneapolis: University of Minneapolis Press.

Lacan, J. (1977) *Écrits: A Selection*, London: Tavistock.

Lafontaine, C. (2009) 'Regenerative medicine's immortal body: from the fight against ageing to the extension of longevity', *Body & Society*, 15, 4: 53–71.

Laqueur, T. (1987) 'Organism, generation, and the politics of reproductive biology', in C. Gallagher and T. Laqueur (eds) *The Making of the Modern Body: Sexuality and Society in the Nineteenth Century*, Berkeley: University of California Press.

Laslett, P. (1989) *A Fresh Map of Life: The Emergence of the Third Age*, London: Wiedenfeld and Nicolson.

Law, J. (2004) *After Method: Mess in Social Science Research*, London: Routledge.

Laws, H. (2005) *Fit to Dance 2: Report of the Second National Inquiry into Dancers' Health and Injury in the UK*, London: Dance UK.

Laz, C. (1998) 'Act your age', *Sociological Forum*, 13, 1: 85–95.

—— (2003) 'Age embodied', *Journal of Aging Studies*, 17, 4: 503–19.

Leder, D. (1990) *The Absent Body*, Chicago: University of Chicago Press.

Lévi-Strauss, C. (1978) *Structural Anthropology*, Harmondsworth: Penguin Books.

Liebling, J. L. (1956) *The Sweet Science*, New York: Viking Press.

Lincoln, Y. S. and Denzin, N. K. (1994) 'The fifth moment', in N. K. Denzin and Y. S. Lincoln (eds) *The Handbook of Qualitative Research*, 3rd ed., Thousand Oaks: Sage Publications.

Lingis, A. (1994) *Foreign Bodies*, London: Routledge.

Lorde, A. (1980) *The Cancer Journals*, Argyle, NY: Spinsters.

Lorde, A. (1988) *A burst of Light*, Athica, NY, Firebrand Books.

Maddox, G. L., Jr. (1964) 'Disengagement theory: a critical evaluation', *The Gerontologist*, 4, 1: 80–82, 103.

Mailer, N. (1975) *The Fight*, London: Penguin Books.

Malbon, B. (1999) *Clubbing: Dancing, Ecstasy and Vitality*, London: Routledge.

Malthus, T. R. (1798) *An Essay on the Principle of Population as It Affects the Future Improvement of Society, with Remarks on the Speculations of Mr. Godwin, M. Condorcet, and Other Writers*, London: T. Malthus.

Manning, P. (1992) *Erving Goffman and Modern Sociology*, Cambridge: Polity Press.

Marcus, G. E. (1994) 'On ideologies of reflexivity in contemporary efforts to remake the human sciences', *Poetics Today*, 15, 3: 383–404.

—— (1998) 'Imagining the whole: ethnography's contemporary efforts to situate itself', in G. E. Marcus *Ethnography through Thick and Thin*, Princeton: Princeton University Press.

Marcus, G. E. and Fischer, M. J. (eds) (1986) *Anthropology as Cultural Critique: An Experimental Moment in the Social Sciences*, Chicago: Chicago University Press.

Marquez, D. X., Bustamante, E., Blissmer, B. J. and Prohaska, T. R. (2009) 'Health promotion for successful aging', *American Journal of Lifestyle Medicine*, 3, 1: 12–19.

Marshall, B. and Katz, S. (2002) 'Forever functional: sexual fitness and the ageing male body', *Body & Society*, 8, 4: 43–70.

Martin, E. (1987) *The Woman in the Body*, Milton Keynes: Open University Press.

—— (1994) *Flexible Bodies: The Role of Immunity in American Culture from the Days of Polio to the Age of Aids*, Boston: Beacon Press.

Marx, K. (1975 [1844]) 'Economic and philosophical manuscripts', in *Karl Marx: Early Writings*, Harmondsworth: Penguin Books, in association with New Left Review.

Mauss, M. (1973 [1935]) 'The techniques of the body,' *Economy and Society*, 2, 1: 70–88.

McCall, M. M. (2000) 'Performance ethnography: a brief history and some advice', in N. K. Denzin and Y. S. Lincoln (eds) *Handbook of Qualitative Research*, 2nd ed., Thousand Oaks: Sage Publications.

McKie, L. and Backett-Milburn, K. (eds) (2001) *Constructing Gendered Bodies*, Basingstoke: Palgrave Macmillan.

MacSween, M. (1993) *Anorexic Bodies: A Feminist and Sociological Perspective on Anorexia*, London: Routledge.

Mead, G. H. (1934) *Mind, Self and Society*, Chicago: Chicago University Press.

Merleau-Ponty, M. (1962) *Phenomenology of Perception*, London: Routledge & Kegan Paul.

—— (1968) *The Primacy of Perception*, Evanston: Northwestern University Press.

Mienczakowski, J. (2001) 'Ethnodrama: performed research – limitations and potential', in P. Atkinson, A. Coffey, S. Delamont, J. Loftland and L. Loftland (eds) *Handbook of Ethnography*, London: Sage Publications.

Miller, D. (1995) 'Introduction: anthropology, modernity and consumption', in D. Miller (ed.) *Worlds Apart: Modernity through the Prism of the Local*, London: Routledge.

Millett, K. (1977) *Sexual Politics*, London: Virago Press.

Minh-Ha, T. (1989) *Woman, Native, Other*, Bloomington: Indiana University Press.

Moi, T. (1985) *Sexual/Textual Politics*, London: Methuen.

Morgan, D. L. and Scott, S. (eds) (1993) *Body Matters*, London: Sage Publications.

Morris, G. (1996) ' "Styles of Flesh": Gender in the Dances of Mark Morris', in G. Morris (ed.) *Moving Words: Re-Writing Dance*, London: Routledge.

Morris, G. (2006) *A Game for Dancers: Performing modernism in the postwar years 1945–1960*, Middletown, Conneticut, Wesleyan University Press.

Mortimer, J. (2001) *The Summer of a Dormouse: A Year of Growing Old Disgracefully*, London: Penguin Books.

Muggleton, D. (2000) *Inside Sub-Culture: The Postmodern Meaning of Style*, Oxford: Berg.

Mulkay, M. G. (1985) *The Word and the World: Explorations in the Form of Sociological Analysis*, London: George Allen and Unwin.

Mulvey, L. (1975) 'Visual pleasure and narrative cinema', *Screen*, 16, 3: 6–18.

—— (1989) 'Afterthoughts on "Visual pleasure and narrative cinema" Inspired by King Vidor's Duel in the Sun', in L. Mulvey *Visual and Other Pleasures*, Basingstoke: Macmillan.

Murphie, A. and Potts, J. (2003) *Culture and Technology*, Basingstoke: Palgrave Macmillan.

Nast, H. J. and Pile, S. (eds) (1998) *Places Through the Body*, London: Routledge.

Nead, L. (1992) *The Female Nude: Art, Obscenity and Sexuality*, London: Routledge.

Needham, R. (ed.) (1973) *Right and Left: Essays on Symbolic Classification*, Chicago: Chicago University Press.

Negrin, L. (2002) 'Cosmetic surgery and the eclipse of identity', *Body & Society* 8, 4: 21–42.

Ness, S. A. (1992) *Body Movement and Culture: Kinesthetic and Visual Symbolism in a Philippine Community*, Philadelphia: University of Pennsylvania Press.

—— (1996) 'Dancing in the field: some notes from memory', in S. L. Foster (ed.) *Corporealities*, London: Routledge.

—— (2004) 'Being a body in a cultural way: understanding the culture in the embodiment of dance', in H. Thomas and J. Ahmed (eds) *Cultural Bodies: Ethnography and Theory*, Oxford: Blackwell Publishers.

Nettleston, S. and Watson, J. (1998) *The Body in Everyday Life*, London: Routledge.

Nicholson, L. J. (ed.) (1990) *Feminism/Postmodernism*, London: Routledge.

Novack, C. J. (1993) 'Ballet, gender and cultural power', in H. Thomas (ed.) *Dance, Gender and Culture*, Basingstoke: Macmillan Press.

Oakley, A. (1974) *Sex, Gender and Society*, London: Temple Smith.
—— (1979) *Becoming a Mother*, Oxford: Martin Robertson.
—— (2007) *Fracture: Adventures of a Broken Body*, Bristol: Polity Press.
Oates, J. C. (1987) *On Boxing*, London: Bloomsbury Publishing Ltd.
Öberg, P. (2003) 'Image versus experience of the aging body', in C. A. Faircloth (ed.) *Aging Bodies: Images and Everyday Experiences*, Walnut Creek: Altamera Press.
O'Connor, B. (1997) 'Safe sets: dance and "communitas" ', in H. Thomas (ed.) *Dance in the City*, Basingstoke: Macmillan Press.
Okely, J. (2007) 'Fieldwork embodied', in C. Shilling (ed.) *Embodying Sociology: Retrospect, Progress and Prospects*, Oxford: Blackwell Publishing Ltd.
O'Neill, J. (1985) *Five Bodies: The Human Shape of Modern Society*, Ithaca, NY: Cornell University Press.
Orbach, S. (1978) *Fat Is a Feminist Issue*, London: Arrow Books.
Orlan (1996) 'Carnal art', in D. McCorquodale (ed.) *This Is My Body . . . This Is My Software . . .*, London: Black Dog Publishing.
—— (2010) 'In retrospect', in S. Donger, S. Shepherd and Orlan (eds), *Orlan: A Hybrid Body of Art Works*, London: Routledge.
Parker, A. and Sedgwick, E. K. (1995) 'Introduction: peformativity and performance', in A. Parker and E. K. Sedgwick (eds) *Performativity and Performance*, London: Routledge.
Parker, R. and Pollock, G. (eds) (1987) *Framing Feminism: Art and the Women's Movement 1970–1985*, London: Pandora Press.
Paulson, S. (2005) 'How various "cultures of fitness" shape subjective experiences of growing older', *Age and Society*, 25: 229–44.
Phelan, P. (1993) *Unmarked: The Politics of Performance*, London: Routledge.
—— (1997) *Mourning Sex: Performing Public Memories*, London: Routledge.
—— (1999) 'Yvonne Rainer: from dance to film', in Y. Rainer (ed.) *A Woman Who . . .: Essays, Interviews, Scripts*, Baltimore: PAJ Books, The Johns Hopkins University Press.
Phillips, D. (1973) *Abandoning Method*, San Francisco: Jossey-Bass Publications.
Phillipson, C. (2007) 'Understanding the baby boom generation: comparative perspectives', *International Journal of Ageing and Later Life*, 2, 2: 7–11.
Pile, S. (1996) *The Body and the City*, London: Routledge.
Pile, S. and Thrift, N. (eds) (1995). *Mapping the Subject: Geographies of Cultural Transformation*, London: Routledge.
Pitts, V. (2003) *In the Flesh: The Cultural Politics of Body Modification*, Basingstoke: Palgrave Macmillan.
Polhemus, T. (1975) 'Social bodies', in J. Benthall and T. Polhemus (eds) *The Body as a Medium of Expression*, London: Allen Lane.
—— (ed.) (1978) *Social Aspects of the Human Body*, Harmondsworth: Penguin Books.

Pollock, G. (1988) *Vision and Difference*, London: Routledge.

Price, J. and Shildrick, M. (eds) (1999) *Feminist Theory and the Body*, Edinburgh: University of Edinburgh Press.

Probyn, E. (2004) 'Eating for a living: a rhizo-ethology of bodies', in H. Thomas and J. Ahmed (eds) *Cultural Bodies: Ethnography and Theory*, Basingstoke: Palgrave Macmillan.

Punday, D. (2003) *Narrative Bodies: Towards a Corporeal Narratology*, Basingstoke: Palgrave Macmillan.

Putman, M. (2002) 'Linking aging theory and disability models: increasing the potential to explore aging with physical impairment', *The Gerontologist*, 42, 6: 799–806.

Rabinow, P. (1986) *The Foucault Reader: an Introduction to Foucault's Thought*, London: Penguin Books.

Rambo Ronai, C. (2000) 'Managing aging in young adulthood: the "aging" table dancer', in J. F. Gubrium and J. A. Holstein (eds) *Aging and Everyday Life*, Malden, MA: Blackwell Publishers Inc.

Rexbey, H. and Povlsen, J. (2007) 'Visual signs of ageing: what are we looking at?', *International Journal of Ageing and Later Life*, 2, 1: 61–83.

Richardson, L. (1997) *Fields of Play: Constructing an Academic Life*, New Brunswick, NJ: Rutgers University Press.

Rose, B. (1993) 'Is It art?: Orlan and the transgressive art', *Art in America*, February: 82–7, 125.

Rose, N. (1996) *Inventing Our Selves: Psychology, Power and Personhood*, Cambridge: Cambridge University Press.

—— (2001) 'The politics of life Itself', *Theory, Culture and Society*, 18, 6: 1–30.

Royce, A. P. (1980) *The Anthropology of Dance*, Bloomington: Indiana University Press.

Russo, M. (1986) 'Female grotesque: carnival and theory', in T. de Lauretis (ed.) *Feminist Studies/Critical* Studies, Basingstoke: Macmillan Press, pp. 213–229.

Saldana, J. (ed.) (2005) *Ethnodrama: An Anthology of Reality Theatre*, Walnut Creek, CA: Altamira Press.

—— (2011) *Ethnotheatre: Research from Page to Stage*, Walnut Creek, CA: Left Coast Press.

Sammons, J. T. (1989) *Beyond the Ring: The Role of Boxing in American Society*, Urbana: University of Illinois Press.

Savigliano, M. E. (1995) *Tango and the Political Economy of Passion*, Boulder: Westview Press.

Schechner, R. (1985) *Between Theater & Anthropology*, Philadelphia: University of Pennsylvania Press.

Scheflen, A. E. (1964) 'The significance of posture in communication systems', *Psychiatry: Journal for the Study of Interpersonal Processes*, 27, 4: 316–31.

Schieffelin, E. L. (1998) 'Problematizing performance', in F. Hughes-Freeland (ed.) *Ritual, Performance, Media*, London: Routledge.
Schneider, R. (1997) *The Explicit Body in Performance*, London: Routledge.
Schultze, L. (1990) 'On the muscle', in J. Gaines and C. Herzog (eds) *Fabrications: Costume and the Female Body*, New York: Routledge.
Schwaiger, E. (2012) *Ageing, Gender, Embodiment and Dance*, Basingstoke: Palgrave Macmillan.
Schwaiger, L. (2006) 'To be forever young? Towards reframing corporeal subjectivity in maturity', *International Journal of Ageing and Later Life*, 1, 1: 11–41.
Seale, C. (1998) *Constructing Death: The Sociology of Dying and Bereavement*, Cambridge: Cambridge University Press.
—— (1999) *The Quality of Qualitative Research*, London: Sage.
Seymour, W. (1998) *Remaking the Body: Rehabilitation and Change*, London: Routledge.
Shakespeare, T. and Watson, N. (2001) 'The social model of disability: an outdated ideology?', *Research in Social Science and Disability*, 1, 2: 9–28.
Sheard, K. G. (1997) 'Aspects of boxing in the Western "civilizing process" ', *International Review of the Sociology of Sport*, 32, 1: 31–37.
Shepherd, S. and Wallis, M. (2004) *Drama/Theatre/Performance*, London: Routledge.
Shildrick, M. (1997) *Leaky Bodies and Boundaries: Feminism, Postmodernism and (Bio) Ethics*, London: Routledge.
Shilling, C. (1993) *The Body and Social Theory*, London: Sage Publications.
—— (2001) 'Embodiment, experience and theory: in defence of the sociological tradition', *The Sociological Review*, 49, 3: 327–44.
—— (ed.) (2007) *Embodying Sociology: Retrospect, Progress and Prospects*, Oxford: Blackwell Publishing.
—— (2008) *Changing Bodies: Habit, Crisis and Creativity*, London: Sage Publications.
Shusterman, R. (2008) *Body Consciousness*, Cambridge: Cambridge University Press.
Simmel, G. (1950 [1908]) 'The stranger', in K. Wolff (ed.) *The Sociology of Georg Simmel*, New York: The Free Press.
Singer, M. (1972) *When a Great Tradition Modernizes*, New York: Praeger Press.
Sklar, D. (2000) 'Reprise: on dance ethnography', *Dance Research Journal*, 32, 1: 70–77.
—— (2001) 'Towards a cross-cultural conversation on bodily knowledge', *Dance Research Journal*, 33, 1: 91–92.
Smart, B. (1985) *Michel Foucault*, Chichester: Ellis Horwood Ltd and London: Tavistock Publications Ltd.

Solomon, R. and Micheli, L. J. (1986) 'Technique as a consideration in modern dance injuries', *The Physician and Sportsmedicine*, 14, 8: 83–90.

Spence, J. (1986) *Putting Myself in the Picture: A Political and Personal Autobiography*, London: Camden Press.

—— (1995) *Cultural Sniping: The Art of Transgression*, London: Routledge.

Stacey, J. (1988) 'Can there be a feminist ethnography?', *Women's Studies International Forum*, 11, 1: 21–27.

Stanley, L. (1990) 'Doing ethnography, writing ethnography: a comment on Hammersley', *Sociology*, 24, 4: 617–27.

Stocking, G. (ed.) (1983) *Observers Observed: Essays in Ethnographic Fieldwork*, Madison: University of Wisconsin Press.

Sudnow, D. (1993 [1978]) *Ways of the Hand: A Re-Written Account*, Cambridge, MA: MIT Press.

Sugden, J. (1996) *Boxing and Society: An International Analysis*, New York: Manchester University Press.

Swingewood, A. (1998) *Cultural Theory and the Problem of Modernity*, Basingstoke: Macmillan.

Sydie, R. A. (1987) *Natural Women, Cultured Men: A Feminist Perspective on Sociological Theory*, Milton Keynes: Open University Press.

Synnott, A. (1993) *The Body Social: Symbolism, Self and Society*, London: Routledge.

Tait, S. (2007) 'Television and the domestication of cosmetic surgery', *Feminist Media Studies*, 7, 2: 119–35.

Tarr, J. and Thomas, H. (2011) 'Mapping embodiment: methodologies for representing pain and injury', *Qualitative Research*, 11, 2: 141–57.

The Boston Women's Health Book Collective (1973) *Our Bodies Ourselves: A Book by and for Women*, New York: Simon and Schuster.

Thomas, H. (1995) *Dance, Modernity and Culture: Explorations in the Sociology of Dance*, London: Routledge.

—— (1996) 'Do you want to join the dance: postmodernism, poststructuralism, the body, and dance', in G. Morris (ed.) *Moving Words: Re-Writing Dance*, London: Routledge.

—— (2000) 'Dance halls: where older people come into visibility in the city', in S. Pile and N. Thrift (eds) *City A-Z*, London: Routledge.

—— (2003) *The Body, Dance and Cultural Theory*, Basingstoke: Palgrave Macmillan.

—— (2004) 'Mimesis and alterity in the African Caribbean quadrille: ethnography meets history', *Cultural and Social History*, 1, 3: 280–301.

Thomas, H. and Miller, N. (1997) 'Ballroom blitz', in H. Thomas (ed.) *Dance in the City*, Basingstoke: Macmillan Press.

Thomas, H. and Walsh, D. F. (1998) 'Modernity/postmodernity', in C. Jenks (ed.) *Core Sociological Dichotomies*, London: Sage Publications.

Thomas, H. and Cooper, L. (2002) 'Dancing into the third age: social dance as cultural text: research in progress', *Dance Research*, 20, 1: 54–80.

—— (2003) *Dancing into the Third Age: Social Dance as Cultural Text: A Research Report*, London: Goldsmiths College.

Thomas, H. and Ahmed, J. (eds) (2004) *Cultural Bodies: Ethnography and Theory*, Oxford: Blackwell.

Thompson, P., Itzin, C. and Abendstern, M. (1990) *I Don't Feel Old: The Experience of Later Life*, Oxford: Oxford University Press.

Thrasher, F. (1927) *The Gang: A Study of 1313 Gangs in Chicago*, Chicago: Chicago University Press.

Thrift, N. (2004) 'Bare life', in H. Thomas and J. Ahmed (eds) *Cultural Bodies: Ethnography and Theory*, Malden: Blackwell Publishing.

Tulle, E. (2008) *Ageing, the Body and Social Change: Running in Later Life* Basingstoke: Palgrave Macmillan.

Turner, B. S. (1984) *The Body and Society*, Oxford: Basil Blackwell.

—— (1991) 'Recent developments in the theory of the body', in M. Featherstone, M. Hepworth and B. S. Turner (eds) *The Body: Social Processes and Cultural Theory*, London: Sage Publications.

—— (1992) *Regulating Bodies: Essays in Medical Sociology*, London: Routledge.

—— (1995) 'Aging and identity: some reflections on the somatization of the body', in M. Featherstone and A. Wernick A. (eds) *Images of Aging: Cultural Representations of Later Life*, London: Routledge.

Turner, B. S. and S. P. Wainwright (2003) 'corps de ballet: the case of the injured ballet dancer, *Sociology of Health and Illness*, 25, 4: 269–88.

Turner, V. (1974) *Dramas, Fields and Metaphors*, Ithaca, NY: Cornell University Press.

—— (1982) *From Ritual to Theatre*, New York: PAJ Publications.

—— (1986) *The Anthropology of Performance*, New York: PAJ Publications.

Turner, V. with Turner, E. (1982) 'Performing ethnography', *The Drama Review*, 26, 2: 33–50.

—— (2007) 'Clothing, age and the body: a critical review', *Ageing and Society*, 27, 2: 285–305.

Updike, J. (1994) 'The disposable rocket', in L. Goldstein (ed.) *The Male Body: Features, Destinies, Exposure*, Ann Arbor: Michigan University Press.

Van Gennep, A. (1960 [1909]) *The Rites of Passage*, Chicago: Chicago University Press.

Van Maanen, J. (1988) *Tales of the Field*, Chicago: University of Chicago Press.

Vincent, J. A. (2003) *Old Age*, London: Routledge.

—— (2006) 'Ageing contested: anti-ageing science and the cultural construction of old age', *Sociology*, 40, 4: 681–98.

Wacquant, L. J. D. (1995) 'Pugs at work: bodily capital and bodily labour among professional boxers', *Body & Society*, 1, 1: 65–94.

—— (1998) 'A fleshpeddler at work: power, pain, and profit in the prize-fighting economy', *Theory and Society*, 27, 1: 1–42.

—— (2004) *Body and Soul: Notebooks of an Apprentice Boxer*, Oxford: Oxford University Press.

Wainwright, S. P. and Turner, B. S. (2003) 'Reflections on embodiment and vulnerability', *Medical Humanities*, 29: 4–7.

—— (2004) 'Narratives of embodiment: body, aging and career in royal ballet dancers', in H. Thomas and J. Ahmed (eds) *Cultural Bodies: Ethnography and Theory*, Oxford: Blackwell.

—— (2006) ' "Just crumbling to bits"? An exploration of the body, ageing, injury and career in classical ballet dancers', *Sociology* 40, 2: 237–55.

Wainwright, S. P., Williams, C. and Turner, B. S. (2005) 'Fractured identities: injury and the balletic body', *Health*, 9, 1: 49–66.

Ward, A. (1993) 'Dancing in the dark: rationalism and the neglect of social dance', in H. Thomas (ed.) *Dance, Gender and Culture*, London: Macmillan Press.

—— (1997) 'Dancing around meaning', in H. Thomas (ed.) *Dance in the City*, Basingstoke: Macmillan Press.

Watson, N. and Cunningham-Burley, S. (eds) (2001) *Reframing the Body*, Basingstoke: Palgrave Macmillan.

Weber, M. (1976 [1905]) *The Protestant Ethic and the Spirit of Capitalism*, London: George Allen and Unwin.

—— (1978 [1921]) 'The history of the piano', in W.G. Runciman (ed.) *Weber: Selections in Translation*, Cambridge: Cambridge University Press.

Weinberg, S. K. and Arond, H. (1952) 'The occupational culture of the boxer', *American Journal of Sociology*, 57, 5: 460–69.

Weiss, G. (1999) *Body Images: Embodiment as Corporeality*, New York: Routledge.

Weiss, G. and Haber, H. F. (eds) (1999) *Perspectives on Embodiment: The Intersection of Nature and Culture*, New York: Routledge.

Welton, D. (ed.) (1998) *Body and Flesh: A Philosophical Reader*, Oxford: Blackwell Publishers.

—— (ed.) (1999) *The Body: Classic and Contemporary Readings*, Oxford: Blackwell Publishers.

Willetts, D. (2010) *The Pinch: How the Baby Boomers Took Their Children's Future – and How They Can Give It Back*, London: Atlantic Books.

Williams, S. J. and Bendelow, G. (1998) *The Lived Body: Sociological Themes, Embodied Issues*, London: Routledge.

Willis, P. (1977) *Learning to Labour: How Working Class Kids Get Working Class Jobs*, Farnborough: Saxon House.
—— (2000) *The Ethnographic Imagination*, Cambridge: Polity Press.
Wilson, S. (1996) 'L'histoire d'O: sacred and profane', in D. McCorquodale (ed.) *Orlan: This Is My Body . . . This Is My Software . . .* London: Black Dog Publishing.
Wolf, N. (1991) *The Beauty Myth*, London: Vintage.
Wolff, J. (2008) *The Aesthetics of Uncertainty*, New York: Columbia University Press.
Wood, C. (2006) *Yvonne Rainer: The Mind Is a Muscle*, London: Afterall Books.
Woodward, Kathleen (2006) 'Performing age, performing gender', *NWSA Journal*, 18, 1: 162–89.
Woodward, Kath. (2008) 'Hanging out and hanging about: insider/outsider research in the sport of boxing', *Ethnography*, 9, 4: 536–61.
Wulff, H. (1988) *Twenty Girls: Growing Up, Ethnicity and Excitement in a South London Micro Culture*, Stockholm: Almqvist & Wiksell International.
—— (2003) 'The Irish body in motion: national identity and dance', in N. Dyck and E. P. Archetti (eds) *Sport, Dance and Embodied Identities*, Oxford: Berg.
Youmans, E. G. (1969) 'Some perspectives on disengagement theory', *The Gerontologist*, 9, 4: 254–58.
Young, I. M. (1998) 'Throwing like a girl', in D. Welton (ed.) *Body and Flesh: A Philosphical Reader*, Malden and Oxford: Blackwell.

ELECTRONIC REFERENCES

ASAPS (2006) Annual Survey: Quick Facts. Online. Available HTTP: http://www.surgery.org/media/news-releases/115-million-cosmetic-procedures-in-2006 (Accessed: 9 November 2012).
ASAPS (2010) Annual Survey. Online. Available HTTP: http://www.surgery.org/sites/default/files/Stats2010_1.pdf (Accessed: 7 November 2012).
ASAPS (2010) Press Release. Online. Available HTTP: http://www.surgery.org/media/news-releases/demand-for-plastic-surgery-rebounds-by-almost-9percent (Accessed: 9 November 2012).
ASAPS (2010) Survey of Views on Plastic Surgery. Online. Available HTTP: http://www.surgery.org/media/news-releases/survey-shows-that-more-than-half-of-americans-approve-of-cosmetic-plastic-surgery (Accessed: 22 May 2012).
BAAPS (2005) Annual Audit. Online. Available HTTP: http://www.baaps.org.uk/about-us/audit/49-over-22000-surigcal-procedures-in-the-uk-in-2005 (Accessed: 18 May 2012).

BAAPS (2010) Annual Audit. Online. Available HTPP: http://www.baaps.org.uk/about-us/audit/854-moobs-and-boobs-double-ddigit-rise (Accessed: 22 May 2012).

Connolly, M.K. and Redding, E. (2010) Dancing towards Well-being in the Third Age: Literature Review on the Impact of Dance on Health and Well-being among Older People, London: Trinity Laban Conservatoire of Music and Dance. Online. Available HTTP: http://www.trinitylaban.ac.uk/media/315435/literature%20review%20impact%20of%20dance%20elderly%20populations%20final%20draft%20with%20logos.pdf (Accessed: 22 May 2012).

Directgov (2012) Calculating Your State Pension Age. Online: Available HTTP: http://www.direct.gov.uk/en/Pensionsandretirementplanning/StatePension/DG_4017919 (Accessed: 22 May 2012).

Frost & Sullivan (2008) Cosmetic Surgery Market: Current Trends. Online. Available HTPP: http://www.frost.com/prod/servlet/market-insight-top.pag?docid=153913646 (Accessed: 7 November 2012).

McVey, R. (n.d.) Elderly Woman in a Yoga Pose. Online. Available HTTP: http://www.gettyimages.co.uk, Creative (RF) #AA023227 (Accessed: 7 November 2012).

Office of National Statistics (2009) Ageing: Fastest Increase in the Oldest Old, Office of National Statistics. Online. Available HTTP: http://www.cardi.ie/publications/ageingfastestincreaseinthe%E2%80%980ldestold%E2%80%99 (Accessed: 10 November 2012).

Orlan Official Website. Online. Available HTTP: http://www.orlan.net (Accessed: 20 May 2012).

Schlessinger, P. (1997) Got to Get This off My Chest: An Interview with Activist and Artist Matuschka. Online. Available HTPP: http://www.matuschka.net/old/interviews/gtgoffchest.html (Accessed: 20 May 2012).

Tothill, D. (n.d.) *Elderly Couple Dancing at a Tea Dance*, Online, Available HTTP: http://www.photofusion.org search number: 1020095.jpg (Accessed: 22 May 2012).

Vallely, P. (2010) Will the Baby-boomers Bankrupt Britain, The Independent: 11. Online. Available HTTP: http://www.independent.co.uk/news/uk/politics/will-the-babyboomers-bankrupt-britain-1936027.html (Accessed: 3 June 2012).

WHO (2002) Active Ageing: A Policy Framework, World Health Organization. Online. Available HTTP: http://whqlibdoc.who.int/hq/2002/who_nmh_nph_02.8.pdf (Accessed: 10 November 2012).

INDEX